radiotoday

guide to HF data on

FT8 & PSK

using WSJT-X and Fldigi

by

Rob Walker, G3ZJQ

radiotoday is an imprint of the Radio Society of Great Britain

This book is a joint publishing project of:

Radio Society of Great Britain
3 Abbey Court, Priory Business Park, Bedford MK44 3WH, United Kingdom
www.rsgb.org.uk

and

ARRL – the national association of Amateur Radio®
225 Main Street, Newington, CT 06111-1400, USA
www.arrl.org

First Printed 2019

Reprinted 2019 with some revisions

Reprinted Digitally 2020 onwards

© Radio Society of Great Britain, 2019. All rights reserved. No part of this publication may be reproduced, stored in a retrieval system, or transmitted, in any form or by any means, electronic, mechanical, photocopying, recording or otherwise, without the prior written permission or the Radio Society of Great Britain.

ISBN: 9781 9101 9370 9

Editing: Mike Richards, G4WNC
Cover design: Kevin Williams, M6CYB
Typography and design: Mark Pressland
Proofing: Neil Whiteside, G4HUN
Production: Mark Allgar, M1MPA

The opinions expressed in this book are those of the author and are not necessarily those of the Radio Society of Great Britain. Whilst the information presented is believed to be correct, the publishers and their agents cannot accept responsibility for consequences arising from any inaccuracies or omissions.

Printed in Great Britain by Page Bros Print of Norwich

Any amendments or updates to this book can be found at:
www.rsgb.org/booksextra

Index

About the Author	iv
1: Introduction	**1**
2: Building a digital modes station	**3**
Transceivers	3
Setting up the computer	5
Setting up the transceiver for digital modes	7
Interfacing transceivers and computers	8
Time servers	10
Setting up WSJT-X software	11
Setting up Fldigi software	14
Summary	17
3: Operating digital modes	**19**
Introduction	19
Digital modes and the UK HF band plans	20
Using WSJT-X in FT8 mode	20
FT8 DXpedition operating mode: Fox and Hounds	33
FT8 Contest Modes	35
Cooperating programs	35
Using Fldigi in BPSK31 mode	38
Tips for using PSK31	41
Summary	41
4: Technical background to the digital modes	**43**
Introduction	43
Digital technology in the analogue world	43
How digital technology interfaces to the analogue world	46
Overview of a digital transmission system	48
Source encoding	49
Error detection	51
Forward Error Correction	53
Convolution encoding	55
Reed Solomon encoding	57
Low Density Parity Check encoding	58
Synchronisation	58
Modulation methods I - analogue	59
Modulation methods II - digital	62
Generating digital modes	65
The radio channel	66
Demodulating the received signal	67
Fourier analysis	67
Source decoding	71
Summary	72
5: The future of HF digital modes	**73**

About the Author

Rob Walker, G3ZJQ

Rob became interested in radio and electronics construction at a very early age. His father built his first TV from a kit in the late 1950s and Rob stood next to him sorting out resistors by value. He was about seven years old at the time and had to stand on a box to see him work. He then spent many fun filled hours 'chassis bashing' and wiring up valves in his teenage years helped and guided by his father. In an odd exchange of roles Rob became licensed in 1970 but his father didn't get his licence until 1971.

He studied physics at Imperial College, London University and graduated in 1972. He went on to do a doctorate in theoretical solid state physics at Sussex University and now boasts BA BSc MBA DPhil FIET CEng ARCS after his name. His research interest at this time was electrical conduction in binary metallic alloys. This led to research work on amorphous silicon solar cells for Plessey, then CAD modelling for integrated circuit design and finally he became a chief engineer managing the circuit research group. Tiring somewhat with the short term research contracts he was managing Rob joined Mars Electronics in 1986 and entered the commercial world. The next nine years were spent managing the design and development of products ranging from small boat radar systems through to coin recognition mechanisms. This covered electronics design from DC to 10GHz. Rob left Mars Electronics in 1995 and became a consulting engineer and project manager.

As part of his development as a consultant Rob gained an MBA by studying at the Open University. He complemented his consultancy work with teaching at the OU as an Associate Lecturer. Slowly he started to do less consultancy work and more teaching and has covered both undergraduate and postgraduate modules in many quite diverse topics including software engineering, technology and project management. Currently, he is engaged as an MSc Supervisor in the Faculty of Mathematics, Computing and Technology at the Open University.

Throughout all this time Rob has remained an active Radio Amateur with interests spanning HF through to UHF.

1

Introduction

Amateur radio continues to develop over the years from its beginnings in the early 20th century to the present. Each time a new activity is spawned there are those that embrace it and those that deem it not in the spirit of the hobby and oppose its creation. Perhaps the biggest change in recent times has been the rise of the internet as both an information resource and a communications medium, but it is not the only influence digital technologies have had upon our hobby. Digital communication technologies are all around us from the PC to the tablet and mobile phone. Today, they are our constant companions. From this milieu have risen the digital modes that this book is all about.

You could argue that the first digital mode was CW using Morse code and a straight key, after all it is an on-off modulation method: early CW transmitters used to key the PA directly! You could argue that the first digital mode was CW using Morse code and a straight key, after all it is an on-off modulation method: early CW transmitters used to key the PA directly! However, Morse code is a human readable system of communication that relies upon highly skilled operators. It is not what is meant today by a digital mode. Radio Teletype (RTTY) was one of the earliest digital modes but it relied upon mechanical teletype machines that were both large and noisy to operate. Today digital modes are intimately tied to computer technology and software. They can be explicitly digital with a computer screen and keyboard or they can be more hidden as with the mobile phone. The power of computers has increased dramatically in recent times and with it so has the number and use of digital modes in amateur radio.

The IBM PC was initially an expensive desktop computer aimed at the business world but, when IBM lost the exclusivity of its product in the 1980s, prices dropped with the competition that ensued. The PC became a home and office product and with the home use came music and entertainment. This was enhanced by the fast developing games market. A market that demanded realism in its gaming environment and this meant a demand for higher computing performance and especially better sound. The PC sound card was spawned in this environment.

Meanwhile, the amateur transceiver market was evolving from thermionic valve based designs of the 1960s and 1970s, such as the classic FT-101 from Yaesu, towards the fully solid state transceivers of the 1980s. The newer designs were more stable and often had a digital frequency dial that made tuning to an exact frequency much easier.

In the late 1990s, Peter Martinez, G3PLX, designed a new digital mode that combined the functionality of the PC with its sound card and the solid state transceiver. This mode was to improve on the performance of RTTY over a noisy HF channel and use commonly available equipment. PSK31 was born. It used phase shift keying of the RF carrier that proved to be more immune to noise and it had a very narrow bandwidth. It was very successful and paved the way for more PC sound card based modes.

In 2001, Professor Joe Taylor, K1JT designed a new digital mode aimed at meteor scatter QSOs using a PC sound card, interface and a VHF transceiver. This was FSK441 and it is still in use today. Professor Taylor is a Nobel Prize-winning astrophysicist from Princeton University, USA who then went on to develop the WSJT–X software package aimed at weak signal applications. The WSJT modes all use state of the art digital communications technology including error control.

The WSJT-X package is freely available to download and is freeware distributed under the Open GNU licence. Whilst FSK441 was the first mode, there are now many more and recently, with the advent of FT8, WSJT-X has become one of the dominant software packages in amateur digital modes. With the release of WSJT-X version 2.0 it has evolved from the relatively exotic meteor scatter and moonbounce activities into main stream HF DXing.

HF digital modes can be categorised in several ways, by their message – is it free format or a fixed format, by their speed – are they slow or fast modes or by their input method – keyboard or speech. In this book, we will be considering HF digital modes that have free or fixed message structures and have keyboard input. We will not cover digital voice or the fast WSJT modes such as MSK144.

In chapter 2 we will discuss how to set up a HF digital station, what equipment you will need and how to configure it. To make this as practical as possible, we will use two different freely available software packages, WSJT-X and Fldigi.

In chapter 3 we will cover how to operate the digital station. How to use the software packages to achieve QSOs using FT8 and PSK31 and Fldigi.

In chapter 4 we discuss the technical aspects of digital modes and explain some of the ways digital modes achieve their impressive performance.

Finally, in chapter 5 we discuss the future of digital modes.

2

Building a digital modes station

In this chapter, we discuss building a digital modes station. Most, if not all, of the components will be present in the average radio amateur's shack already. The main components are a transceiver, complete with power supply and antenna, and a computer. A simple interface between the two might also be needed.

Transceivers

The primary requirements for a transceiver to be capable of using digital modes are the same as those for SSB and CW: frequency stability and an output transmission low in harmonics. This means that most if not all transceivers produced in the last 20 years are acceptable, providing the older ones still meet their original specifications regarding the two primary requirements. Power output is less important for digital modes than for SSB or CW, so even low power transceivers can be used to good effect. Sophisticated digital processing is also of less importance since speech processing and noise reduction functions are turned off when in digital mode. It is perfectly possible to build your own transceiver and use it rather than buying a commercial product. Digital modes have not disrupted the amateur's ability to do homebrew.

The simplest and most universal method of interfacing to your transceiver to use digital modes is through the microphone input, PTT and headphone output connectors, which are often on the front panel. All transceivers will have these connections. A direct connection is possible, but it is better to provide some form of interface and this will be discussed later.

As computers evolved, their connectivity has improved and it is the same with transceivers. Older transceivers just have Accessory sockets on the back panel that allow audio in and out as well as the PTT function. The link is basically analogue audio and a simple PTT switching voltage. They are usually round DIN type connectors. Serial ports using 9 pin D connectors became common, but these were used mainly for computer control of the transceiver (CAT) and not analogue sound in and out. Finally, transceivers began using the Universal Serial Bus (USB – not to be confused with upper side band!).

Current transceivers, like those in **Figure 1** are especially easy to use in digital modes but they often have an assortment of rear connections that can be confusing.

The terminology for these connections can be confusing because they are called by different names and have slightly different functionality: Computer

Figure 1: Modern transceivers.

2: Building a digital modes station

Aided Transceiver (CAT), Accessory Socket sometimes abbreviated to ACC, Data Socket, USB port and Com port just to name a few. It is necessary to read the User Manuals, and sometimes even the Service Manuals, to gain an understanding of which connection does what and how to connect to it.

A quick glance at the back panel of any modern transceiver will show you what is available. For example, **Figure 2** shows the back panel of the FT-991 and you can see that there are several possibilities for connection: RTTY/ DATA, GPS/CAT and USB as well as an Ext Speaker connection. You might assume that the RTTY/DATA would be the most appropriate, but this turns out to be a connection dedicated to RTTY and not to data in general. The GPS/CAT connection allows computer aided control of the transceiver including PTT and this might be useful for data modes. Since the latest type of connection is USB it is well worth investigating. Both the FT-991 and the IC-7300 feature USB ports that provide both the sound card and the CAT serial control link over a single cable. However, this is not always obvious from the user manuals. Using the transceiver's USB port connection may require you to download specific USB drivers associated with the model of transceiver and install these on your computer. Once you have done this you can directly access the sound card and CAT serial link within the transceiver and no other interface or connection is required.

In summary, basically there are three potential ways of connecting your transceiver to your computer: via the microphone and headphone sockets, via accessory sockets on the back panel and via USB. All but the USB socket connection will require an additional interface.

Setting up the computer

The modern amateur station will undoubtedly have some form of computer and typically this will be a PC either a desktop model or, perhaps more likely, a notebook. It might be a Windows machine, a MAC or a Linux based computer. Any of these will do to establish a digital modes station, but here we

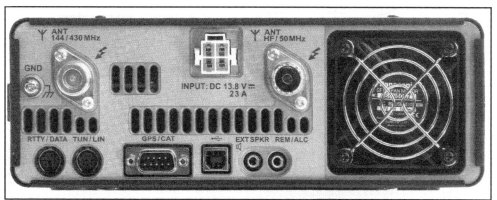

Figure 2: FT-991 back panel.

will concentrate upon Windows as this is probably more common and also setting up under the other operating systems is very similar in concept with only a few differences in procedures. Surprisingly, despite the complexity of the software, it is not necessary to have a powerful computer. As can be seen from the specification to run WSJT software below, the requirements are modest and even older computers can be pressed into service.

- Operating system – Windows (XP or later), Linux or OS X
- 1.5GHz CPU or faster
- 200Mb of memory
- Monitor with at least 1024x780 resolution
- Audio system with a sampling rate of 48000 samples per second at 16 bit resolution

The audio system does not need to be an expensive state of the art device since most internal sound systems will work fine and, failing an internal solution, a cheap external USB sound card will be just as good. If you are using a sound system that is part of your transceiver you will need special drivers. These are available from the manufacturer as a free download.

The Windows 10 PC sound system is configured using the Sound control panel that is found under the Settings tab, itself found by right clicking the bottom left Windows logon button.

Figure 3 shows the sound configuration screen. This has three tabs of interest: Playback, Recording and Sounds. The Playback tab shows the current configuration of the speaker output; here it shows that the USB Audio Device is selected. If you are using an internal sound device then this screen will only show that device since in

Figure 3: Windows Sound Control panel.

2: Building a digital modes station

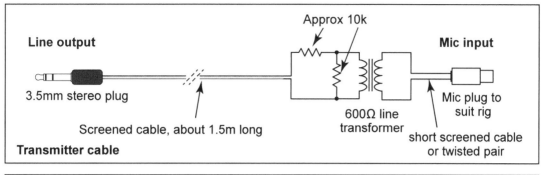

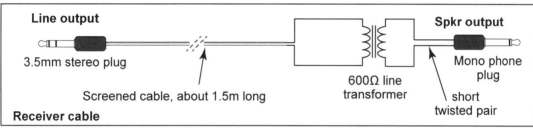

Figure 5: Simple interface circuit.

It can be constructed by various techniques and the layout is not critical, but you should take care to both maintain the isolation and screen the interface to prevent RF interference. The use of ferrite clip-on cores on the input and output leads can help reduce such interference. The simple circuit does not include a PTT connection, but this can either done using the transceiver's VOX or by activating a simple transistor switch from the audio line output. Several designs are available on the internet.

An alternative is to purchase a commercial interface box. An example is shown in **Figure 6**. They often include a separate sound card and are connected using USB.

Finally, if you do not wish to use your computer's sound system then purchase a cheap

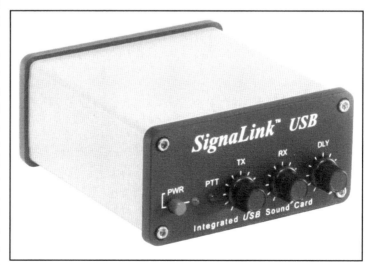

Figure 6: Signal Link USB interface from Tigertronics.

2: Building a digital modes station

the case shown in Figure 3 there are two sound devices an internal device, Realtek, and an external USB Audio device so there is a choice to be made. Make sure you select the one connected to your radio! If you only have one sound device, it is important to prevent Windows from sending its system sounds to your radio, so it is advisable to select 'no sound' under the Sounds tab also found on this screen. If you have a dedicated sound card for digital mode operation, make sure it is not selected as the Windows default playback device but is only selected by your digital modes operating software.

If you double click on the device in the list further configuration options are available. Here you can select the audio sampling rate and bit depth, for example, WSJT-X software uses 16bit and 48000Hz. It is worth checking that this is set up correctly here as well: see **Figure 4**.

Normally you will not need to install a new driver for your sound device. You either use the one that is already configured in your PC or the one automatically loaded when you plug in your USB sound device. However, if you are using a modern transceiver with a built-in sound card, then you will most likely need to install a special driver specific to that transceiver. Consult the User Guide for your specific transceiver to see how to do this. Usually this just involves downloading the driver's installation setup file from the manufacturer's website and running this to install the driver.

Setting up the transceiver for digital modes

The most direct way to set up digital modes is to use an external interface connected to the microphone input, headphone output and PTT on your transceiver. Here you measure and adjust the input and output levels from your computer and radio; you have total control. However, modern transceivers will offer you several other options. They often have a socket on the rear panel that allows audio in/out and PTT that is dedicated to data modes. As explained earlier, unfortunately, there is no common terminology for these sockets, so it is a matter of reading the radio's manual and, to make matters even worse, often

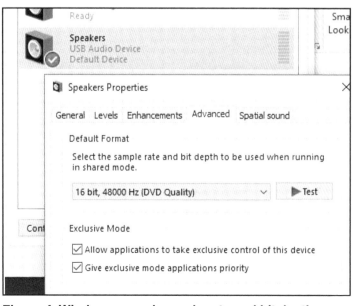

Figure 4: Windows sound sample rate and bit depth.

the manuals are not very clear in this respect. It is worth investigating internet forums for both your radio and for the digital mode software that you are attempting to use. For WSJT-X users https://groups.io/g/WSJTX is a good starting place as is https://groups.io/g/winfldigi for Fldigi software users.

Having an idea of what you are trying to achieve when you set up your transceiver does help. The setup can be viewed as a two-stage process: setting up the radio's communication channel and setting up the radio's RF performance. In modern transceivers, this is done using the menu structure pertinent to the radio in question. Each manufacturer has his own terminology so if you move between major manufacturers you will need to learn the new terms.

Table 1 shows the menu parameters that need to be adjusted on a Yaesu FT-991 transceiver for FT8. In addition to these menu settings, the mode needs to be selected: MODE DATA-USB. Several of the receiver options are also required to be adjusted: DNR is OFF, DNF is OFF. On the transmit side DTGAIN is adjusted to avoid over-driving the PA.

Interfacing transceivers and computers

Unless you are using your transceiver's internal sound card, connected via USB, you will need a simple interface between computer and transceiver. The purpose is to isolate the two devices and reduce the possibility of feedback loops or even damage to either device. A simple interface circuit is shown in **Figure 5.**

Menu			
031	CAT rate	4800	
032	CAT TOT	100mS	
033	CAT RTS	Enable	
060	PC keying PTT	Off	
062	Data Mode	OTHERS	
064	OTHER DISP	1500Hz	This affects the dial frequency vs actual frequency
065	OTHER SHIFT	1500Hz	
070	DATA IN SELECT	REAR	
071	DATA PTT SELECT	RTS	
072	DATA PORT SELECT	USB	

Table 1: Typical menu parameters for FT8 on Yaesu FT891

USB sound card such as the one in **Figure 7**. These are available on the internet for less than £10.

Time servers

As we will see later, it is important to set the computer's real time clock to an accuracy of a few tenths of a second. This is best achieved using a time server. A free solution is to download and install a Network Time Protocol server such as Meinberg: https://www.meinbergglobal.com/english/info/ntp.htm

Figure 7: An external USB sound card powered by the USB connection.

Although this was originally a Unix application, it is now available for Windows and is easily installed using a wizard.

Although you can get some idea of the accuracy of your real time clock from the command line interface using Meinberg, another useful web resource is Time.is, as shown in **Figure 8**.

Figure 8: Checking your time.

2: Building a digital modes station

Setting up WSJT-X software

The WSJT-X software is a comprehensive application that allows communication on all the weak signal modes: FT8, JT4, JT9, JT65, QRA64, ISCAT, MSK144 as well as the weak signal propagation reporter (WSPR) and Echo, a mode specific to moonbounce. The current version at the time of writing (December 2018) is WSJT-X 2.0 and it is available as a free download from: https://physics.princeton.edu/pulsar/k1jt/wsjtx.html

There are versions for Windows, Linux and OS X so you just need to select the one you need. Version 2.0 is a major upgrade from previous versions and since the FT8 protocol has changed it is not backwards compatible with previous versions of WSJT-X when running FT8 (or MK144 but we are not covering that mode here). The windows installation is easy using the usual installation wizard as shown in **Figure 9**.

You just run the downloaded file and follow the prompts. On running the software you should see the main screen and the waterfall.

You must now configure the software so click File on the menu bar at the top of the main screen then click on Settings or you could use the F2 speed key. Start by selecting the Gen-

Figure 9: WSJT-X setup.

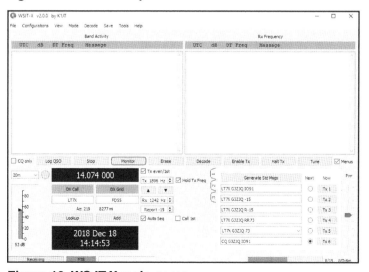

Figure 10: WSJT-X main screen.

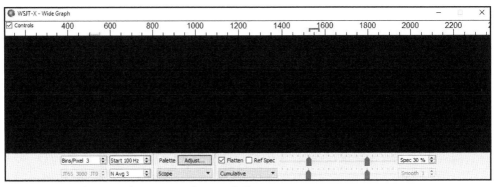

Figure 11: WSJT-X Wide graph / waterfall.

eral Tab, **Figure 12**, and enter your callsign and Maidenhead locator. Next select Radio and enter your radio model using the pull down list, set the COM port, and the appropriate serial port settings for your radio's interface.

Finally, make sure you have the correct audio card/codec selected. Go to the audio Tab, see **Figure 14**, and use the pull down menus to select the correct codec.

Select FT8 from the mode pull down menu at the top of the screen. If you have setup the interface, have accurate network time and are tuned to an FT8 frequency, the waterfall and main screen will slowly become populated with the callsigns of any active stations. If this does not happen you will need to re-check your settings but do remember that it will take a minute or so for the decoded traffic to appear. If after waiting a few minutes nothing has appeared, then check your PC clock against the time service Time.is. You need to be within a second or so of the exact time. There is also an extensive

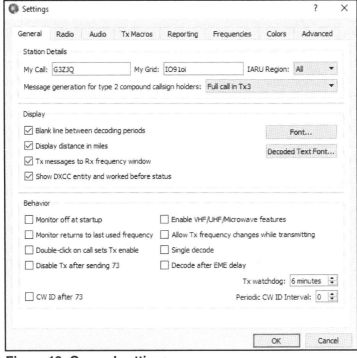

Figure 12: General settings.

2: Building a digital modes station

help file facility accessed by pressing F1 and a User Guide available from https://physics.princeton.edu/pulsar/k1jt/wsjtx-doc/wsjtx-main-2.0.0.html. The WSJT group mentioned earlier is also a good place to get help.

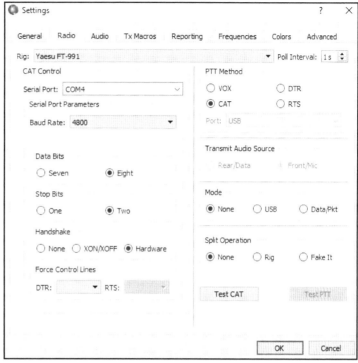

Figure 13: Radio settings.

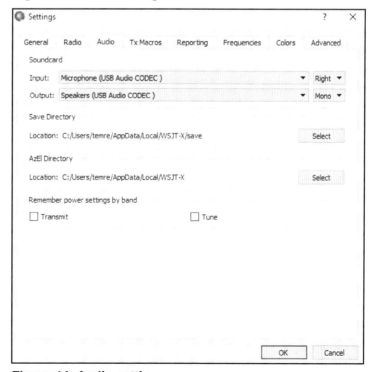

Figure 14: Audio settings.

Setting up Fldigi software

Although WSJT-X is a very comprehensive software package, it only covers the WSJT weak signal modes. If you want to operate using any other of the myriad HF digital modes, then you will need to install more software. Fldigi, which is an acronym for Fast Light Digital modem application, is a comprehensive open source, freeware package that is available for download from Source-Forge at https://sourceforge.net/projects/fldigi/. It is available for Windows, Linux, Mac OS and even Android. The minimum recommended CPU speed is 1.2GHz – 1.6GHz, which is not much of a limitation with modern computers. The Windows version will run under any version of Windows after Win 2000.

The current Windows version is ver 4.0.17 and if you download the setup file and run it then you will see the following opening screen.

There are four windows. The white top left window is the browser window and it will show activity across the full 3kHz audio bandwidth. The yellow window is the received window and it is here that the decoded messages will be displayed. The light cyan window is the transmit window. It is here

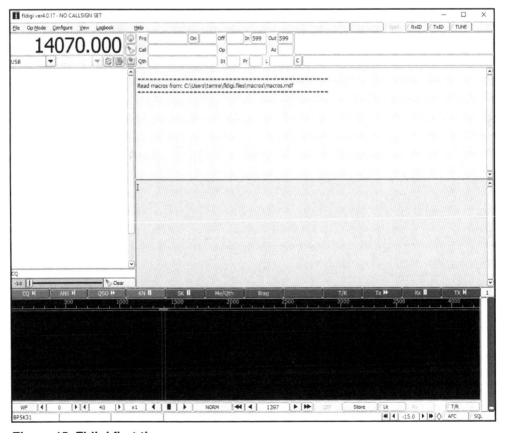

Figure 15: Fldigi first time screen.

2: Building a digital modes station

that you will type your messages to be sent. The lower black window is the waterfall. The two thin vertical lines on the waterfall indicate the frequency that you are tuned to: receiving and transmitting. There are three command ribbons. The top one has the usual File tab followed by OpMode, Configure, View, Logbook and Help. The second command ribbon beginning with CQ is a way of accessing the macros, these are common text messages that you create. The lowest command ribbon controls the waterfall.

Before you can use the software, it must be configured. Configuration is in several stages:
- Enter your stations details
- Set up the audio card
- Set up the CAT control of your transceiver
- Set up the waterfall
- Set up the TX audio level

You can open the station details configuration page by *Configure->UI->Operator.*

You just enter your details in the page as shown in **Figure 16**.

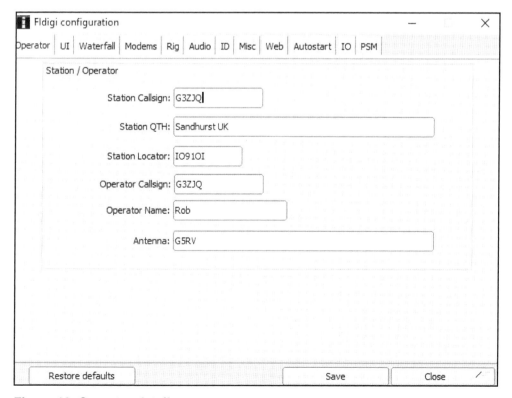

Figure 16: Operator details.

Next, we setup the audio: Configure->Sound Card. This is shown in **Figure 17**.

This is setup with the same information as that used in the WSJT-X setup since the rig is the same. CAT control is next and here Fldigi has several options but since this is to work with the same transceiver as WSJT we can use the same CAT techniques as previously: as shown in Figure18 under the Hamlib tab. We get to the configuration page by Configure->Rig Control

Notice that the selections are the same as the WSJT-X configuration and include the COM port, Baud rate, stop bits, flow control and PTT selection.

Even now that the software is receiving audio from the transceiver the waterfall is probably blank. The initial dynamic range of the waterfall is set from 0dB with a range of 40dB. Since we are using 16bit audio we can adjust the range to 90dB: this is done in the second box from the left in the command ribbon beneath the waterfall. You can now adjust the bottom level until the waterfall noise floor appears. If you now select the mode by OpMode->BPSK31 and choose a frequency where BPSK31 is to be expected, you should now see something like **Figure 19.**

Figure 17: Sound card configuration.

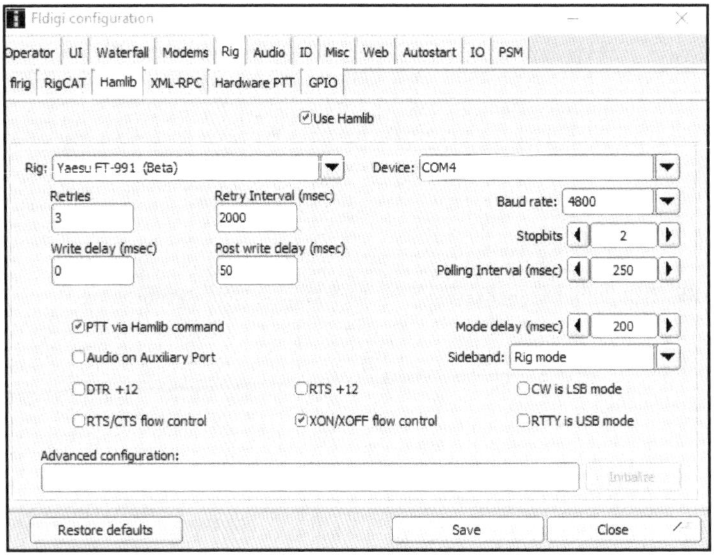

Figure 18: Rig control configuration.

2: Building a digital modes station

Summary

In this chapter we have discussed building a station to operate using digital modes. We discussed transceivers and found that most of them could be pressed into digital mode service without modification. The key requirement of a computer was discussed at some length and we concluded that the requirements were minimal and that most shack computers could be used. We discussed the options concerning interfacing computers to transceivers and offered some simple solutions. Finally, we covered the installation and basic configuration of two different but complimentary software packages: WSJT-X and Fldigi.

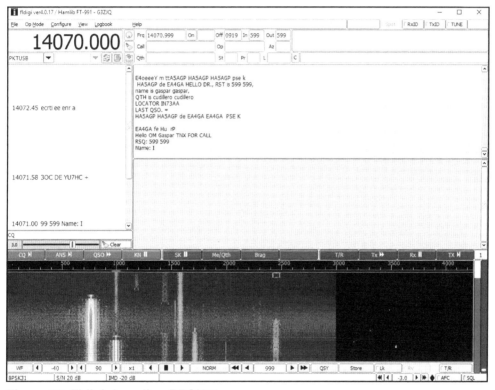

Figure 19: Fldigi decoding BPSK31.

3

Operating digital modes

Introduction

Having set up your transceiver, connected the interface and loaded the software package, you are very nearly ready to get going with your first QSO. We covered the installation and basic configuration of two well-known software packages, WSJT-X and Fldigi, in the previous chapter. In this chapter we will see how to use these packages and enter the world of HF digital. As with many software packages, the supplied instructions can be confusing for those that are new to data modes. It is for this reason that we will go into some detail, in the hope that this might help overcome the confusion. However, the software packages will change over time and they may alter their user interface from that shown here. Hopefully this will not be too different!

It is worth mentioning that you can achieve very good results, sometimes even surprising results, using modest antennas and low power. The power duty-cycle using digital modes is much larger than that of SSB and, as such, it is well worth keeping the power level down to avoid overloading the PA. Most transceivers have a power output rating that is intended for SSB operation and thus should be reduced for high duty-cycle digital modes.

This chapter starts by considering where in the amateur HF bands you can find digital modes being used. It then discusses in some detail how to operate the WSJT-X software using FT8 as an example mode. We then discuss operating a keyboard mode using Fldigi and BPSK31.

Digital modes and the UK HF band plans

Although digital modes can be operated throughout most of the amateur bands the band plans do specify frequency allocations specifically for these modes. In the UK, digital modes are to be found together with other narrow bandwidth modes in the bottom part of the allocated band. **Table 2** shows the current (2018) allocations for the digital modes together with spot frequencies where certain modes can often be found. The PSK frequencies cover BPSK31 as well as other PSK variants.

Using WSJT-X in FT8 mode

Now that you have completed the basic configuration and WSJT-X is working, you can explore the different modes. Let's start with FT8, currently the most popular mode. If you have followed the previous sections configuration and you are tuned to an active FT8 frequency, the waterfall should look something like **Figure 20**, which shows the waterfall and **Figure 21**, which shows the main decode window. Note, these are separate windows and can be moved around independently. If you can hear the characteristic FT8 sounds coming from your transceiver, but no signals appear on the

Frequency band	UK plan		PSK	JT65	FT8
160m	1.830 – 1.840	500Hz BW	1.838	1.838	1.840
80m	3.570 – 3.580 3.580 - 3.590	200Hz BW 500Hz BW	3.570	3.570	3.573
40m	7.040 – 7.047 7.060 – 7.100	500Hz BW 2.7kHz BW	7.040	7.076	7.074
30m	10.130 – 10.150	500Hz BW	10.141	10.138	10.136
20m	14.070 – 14.089	500Hz BW	14.070	14.076	14.074
17m	18.095 – 18.105	500Hz BW	18.098	18.102	18.100
15m	21.070 – 21.090	500Hz BW	21.070	21.076	21.074
12m	24.915 – 24.929	500Hz BW	24.920	24.917	24.915
10m	28.070 – 28.120 28.150 – 28.190	500Hz BW 500Hz BW	28.070 28.120	28.076	28.074
6m	50.300 – 50.400	2.7kHz BW	50.305	50.310	50.313

Table 2: Digital modes frequencies.

3: Operating digital modes

waterfall, check that the monitor button is green, and your computer clock is accurate using Time.is.

The waterfall in Figure 20 shows several FT8 signals spread out over the frequency range of the audio from your receiver and the black box on the main window shows the frequency the transceiver is set at. Since we are set to USB-DATA (or just USB if you are using the microphone input) the actual frequencies of the FT8 transmissions are the sum of the frequency in the box

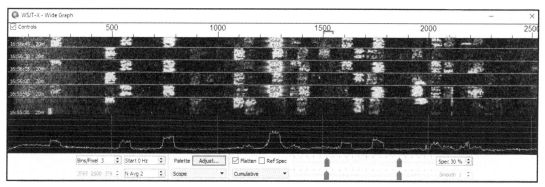

Figure 20: WSJT-X wide graph or waterfall showing several FT8 signals on 20m.

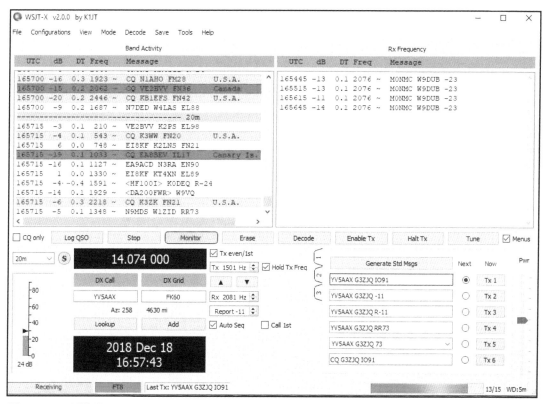

Figure 21: WSJT-X main window.

21

on the main window, 14.074, and the scale along the top of the waterfall. The band activity box shows all the decoded FT8 signals within the audio passband, while the RX Frequency box shows just those within the Rx window defined by the green U-shaped cursor that is on the waterfall scale. The red U-shaped cursor shows the transmit frequency, again on the frequency scale. You can move these cursors around by just clicking on the waterfall to move the Rx cursor, Shift-click to move the Tx cursor and Ctrl-click to move both. These and other special mouse commands can be seen by pressing F5. The waterfall at 770 clearly shows the 15s odd and even periodicity of the FT8 signal. You can identify the station by comparing the frequency given in the waterfall to that in the decoded Band activity box or by placing the receive cursor over the frequency of interest. The individual time periods are listed to the left next to the band in use. The overall scale of the waterfall is governed in the horizontal scale by the Bins/Pixel, Start frequency and by the N Avg in the vertical (time) scale. You can also stretch the window itself. The graph beneath the waterfall, the spectrum window, shows the signal strength in dB.

The main window shows the Band activity and Rx activity boxes. Entries are colour coded to highlight CQ calls and other attributes listed under File-Settings-Colours. Each entry gives the UTC time, signal strength, time delta from your clock, frequency and the message itself. Messages in FT8 are very structured and give basic QSO information: callsign, locator, signal report and acknowledgement. One final adjustment on the receive side is to optimise the receive levels for optimum FT8 decoding. This is done by adjusting the audio input to the sound card to make the WSJT level indicator range between 30-50dB. This can be done in several ways: within Windows, within your transceiver or perhaps even using your interface. The aim is not to have any one adjustment set to maximum so that linearity and distortion is avoided. In Windows 10 you can find the sound control panel under Settings-System-Sound.

Figure 22 is a composite screen shot showing the Windows sound control panel overlaid on the WSJT main window. It shows the microphone input level set to a midrange value of 52 and the WSJT level varying around 50dB. Notice that the active sound device is the USB Codec since I am using an external sound card. If I was using

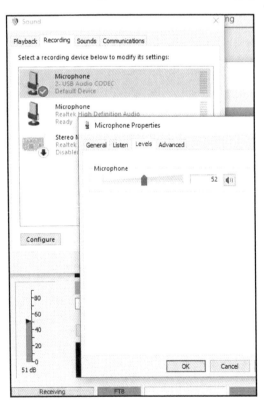
Figure 22: Windows 10 input sound levels.

3: Operating digital modes

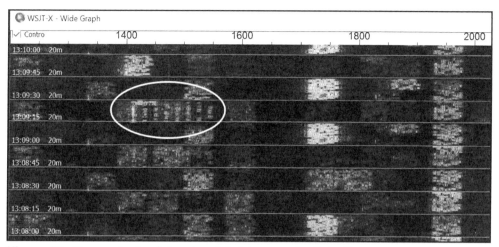

Figure 23: Transmission showing third harmonic distortion.

the internal card and an external interface to the transceiver's microphone socket, then it would be the Realtek microphone that was active and being adjusted. If you are using a data audio output socket on your rig, there may also be an adjustment within the transceiver to consider; this is usually accessed via a menu setting: Data Output Level. Set this to midrange as well.

Before moving on to a live QSO, now is a good time to set the transmit power level. This is done by adjusting the audio input to the transceiver. It is important to set the transmit levels such that the audio frequency part of the transmit chain is operating linearly. If you do not do this correctly, you are in danger of producing distortion products. **Figure 23** shows a partial screen shot taken from the waterfall of an FT8 3rd harmonic distortion product this can be recognised by its bandwidth: it is 150Hz wide rather than the 50Hz of the normal transmission. Note that this is a harmonic of the FT8 audio modulation rather than of the RF signal so, for example, if you are transmitting at 500Hz within the wide graph waterfall this harmonic will be at 1500Hz (and it is likely that the second harmonic will also be present at 1000Hz).

To avoid producing harmonics you must adjust the audio level. However, this is not a single control but can be three or more independent but sequential controls. A block diagram of the audio chain is shown in **Figure 24**. The sound card is shown as a dotted box since it can be either within the

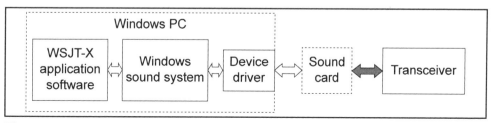

Figure 24: Audio chain.

computer or within the transceiver.

This figure might look strange since it only refers to software blocks and the relationship to the audio is hidden. This is because most of the audio chain is digital and hence software controlled. Only the interface between the Sound card and the transceiver is analogue and this can be inside the transceiver or represent an interface outside of the transceiver. There is an audio level software slider on the right-hand side of the main WSJT-X screen marked Pwr, Windows has a complete audio system that has another software slider that controls the audio output of the sound card and, finally, there is an audio level control within the transceiver that may or may not be software controlled. If you are using an analogue audio input and interface you might also have a further audio level control there as well! So how do you adjust all of these controls? I suggest the following procedure.

1. Adjust the Pwr slider in WSJT-X main screen to midrange
2. Adjust the audio output level (speakers) to midrange. In Windows 10 it is found in the same place as the microphone level adjustment explained previously.
3. If your transceiver has an output power setting/ adjustment set this to a few watts. I use a 10watt setting during setup.
4. Connect your transceiver to a dummy load via a power meter/VSWR bridge
5. Click on the Tune button in WSJT-X and monitor the power output of your transceiver.
   a. If the power output as measured by the power meter matches what you have set in (3) decrease the audio level control within your transceiver until it starts to decrease the output power.
   b. If the power output as measured by the power meter is less than what you have set in (3) increase the audio level control within your transceiver until it just matches the set power.
6. Monitor the ALC level on your transceiver whilst you complete (5). It should not be excessive

This procedure works most of the time, but a better procedure would be to monitor your actual transmission. This is easily accomplished if you have a second independent receiver such as a cheap SDR dongle and some free SDR software. You will need to avoid overloading this monitoring receiver, so a well-defined sampler is required. This need not be complex and is easily constructed from a metal box, three RF connectors and a length of wire. There are many designs for RF samplers available on the internet or in books. Some are quite complex constructions requiring metal working facilities but, for our purposes, a simple construction is adequate. **Figure 25** shows a very simple design that can be constructed from three connectors and a small box. It does not maintain 50Ω throughout, but this is of less importance at

3: Operating digital modes

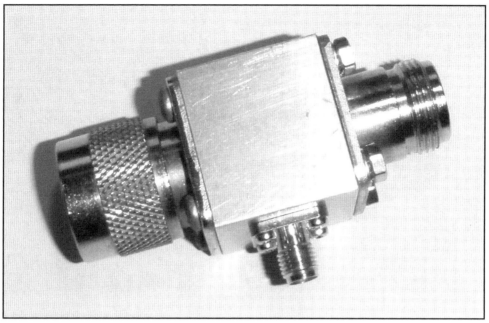

Figure 25: HF Sampler.

HF. Note, the SMA connector is not connected to anything on the inside of the box and just uses the centre connection as a sniffer antenna. The two N connectors are linked by a short straight-through wire link.

At 20m, it has a sampling loss of 80dB but this was not enough to ensure that the monitoring receiver was not overloaded so I added an additional 20dB attenuator. It is important to ensure that you do not overload your monitoring receiver since overloading may generate internal harmonics that would give you a false impression of what would otherwise be a clean signal. The final monitoring setup is shown in **Figure 26**. For initial setting up, I used a 50Ω dummy load but for long-term, this can be replaced by an antenna.

It is easiest to have a completely independent monitoring setup so you will need a second computer to run the SDR software. Tune the monitor-

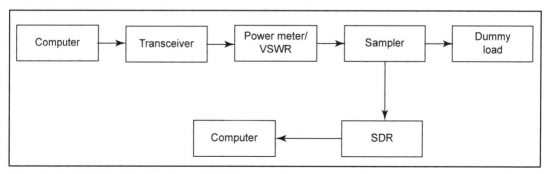

Figure 26: Monitoring setup.

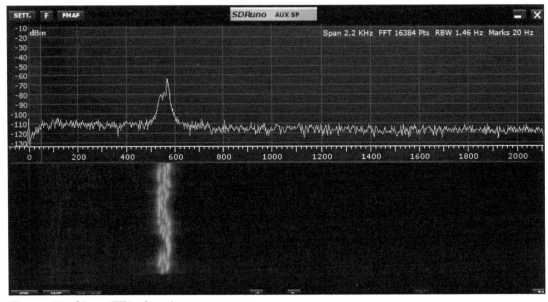

Figure 27: Clean FT8 signal.

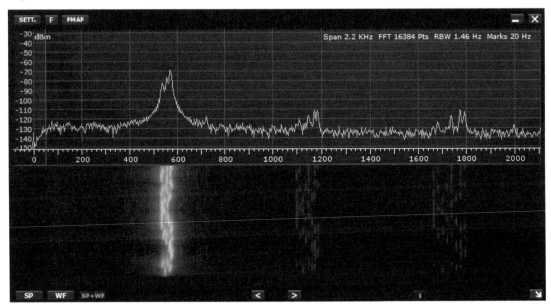

Figure 28: Overdriven FT8 signal showing harmonics.

ing SDR to your transmitted signal and use the spectrum display within the SDR software to examine your signal. A clean FT8 signal without harmonics is shown in **Figure 27.** whilst **Figure 28** shows the harmonics caused by overdriving the audio stages.

You can now follow the previous adjustment procedure whilst actually

3: Operating digital modes

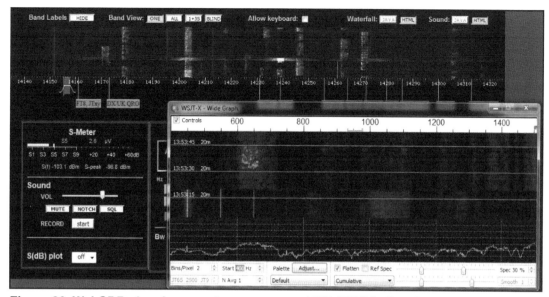

Figure 29: WebSDR showing spectrum scope and WSJT Wide Graph as an insert.

monitoring for any harmonic distortion.

If you do not have access to a suitable SDR receiver you can use a remote WebSDR. This approach will require less equipment, but you will need a second computer linked to the internet. There are several WebSDR that can be used, and you might try several before a suitable one is found. Remember it must be able to receive your signals, so the choice is determined by time of day, distance and operating band. Most WebSDRs do not have a spectrum scope that is detailed enough to show your FT8 signal in sufficient detail to look for harmonics. So just receiving your signal is not enough and you will need to decode the FT8 as well. To do this, you need to link the WebSDR output from your browser to a decoding program such as WSJT-X. It is possible to use the speaker output from your PC to play the SDR output from your browser and a microphone input to the same PC that is connected to the input of WSJT-X but I recommend the use of a virtual cable instead. A virtual cable is a piece of software that allows you to connect the audio output of one application directly to the input of another application running on the same computer. Several virtual cables are available for download from the internet. Once loaded, a virtual cable appears as an audio device within Windows and can be selected as an input or output in the same settings boxes as any other audio device (speakers, microphones etc). **Figure 29** shows a WebSDR monitoring 20m.

The small window marker at 14.155MHz marks the audio pass band that is shown on the WSJT-X Wide Graph. This was taken using the speaker – microphone method discussed earlier and you can see some odd artefacts,

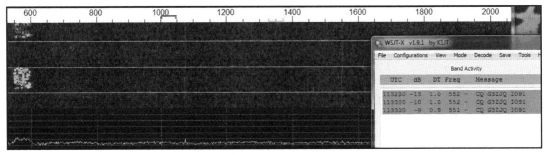

Figure 30: My FT8 signal as received by a WebSDR.

the dark vertical stripes, in the Wide Graph as a result. You can still see my FT8 signal on this graph, but it is quite weak. Using the virtual cable to connect my web browser to WSJT-X directly results in a cleaner signal. This is shown in **Figure 30**.

The absence of any harmonics is confirmation that the audio level has been set correctly.

Of the three methods to set the correct audio level, the sampling method is the most accurate and is recommended. However, using a WebSDR uses less equipment and does give you confidence that your signals are clean. You can use the setup procedure without monitoring, but this is a last resort and not recommended.

Finally, perhaps the cheapest method is to get a friendly local FT8 operator to give you honest feedback on your signal when transmitting on a clear frequency away from the FT8 'watering holes'.

The WSJT-X application has a setting that minimises the transmission of harmonics and this is worth using. Split operation adjusts the transmit frequency to keep the FT8 tones between 1500Hz and 2000Hz that makes any harmonics, which will be greater than 3000Hz, outside the pass band of

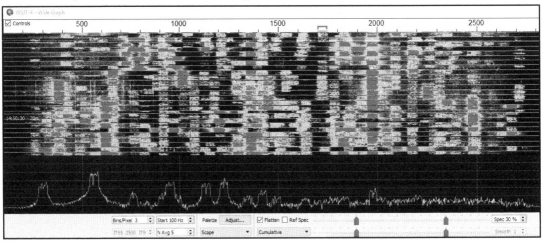

Figure 31: 20m at a busy time!

3: Operating digital modes

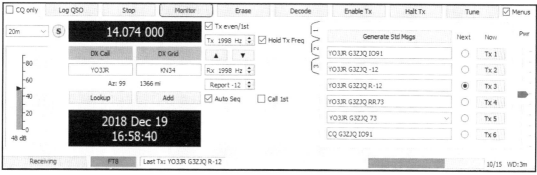

Figure 32: WSJT-X operating window.

the normal SSB transceiver. There are two options to do this: using the transceivers A and B VFOs or letting WSJT-X emulate this by shifting the transmit frequency. For example, if you are transmitting at 550Hz on the waterfall and your transceiver is set to 14.074MHz then your actual transmit frequency will be 14.074550MHz and the potential 2nd harmonic will be 1100Hz. Under split operation when WSJT goes into transmit it will adjust the frequency on the transceiver to 14.073MHz and the FT8 tones to 1550Hz. The actual transmit frequency is still 14.074550MHz but now the second harmonic will be at 3100Hz which is above the pass band and hence it will be attenuated. Split operation is set up under File->settings->radio. If you chose **rig** then WSJT-X will use the VFO A/B method and switch between VFOs when you transmit. Make sure that you have set up both VFOs to data mode otherwise you may find yourself attempting to transmit on USB or even LSB!

Now the receive and transmit setup is complete, we can start making QSOs. Firstly, let's consider how to set the transmit frequency. WSJT show you all the activity in a 3000Hz bandwidth and multiple stations can easily fit into this range, since FT8 only occupies 50Hz. On a typically busy day on 20m this can look crowded, **Figure 31**.

The aim is to choose a frequency that is clear; these are dark blue on the waterfall. It is not necessary or desirable to call a station on its own frequency since split frequency operation is preferable (note this is different from split operation discussed earlier). It avoids the pile-up when a station calling CQ has multiple respondents calling him each causing QRM to the others. However, the choice of transmit frequency can only be done by selecting a clear frequency at the transmitting station's end and this could cause QRM at the DX stations location or indeed elsewhere. The FT8 software is tolerant of overlapping stations and this helps in this situation.

Once you have chosen a clear frequency there are a few other settings that you might find useful before you start transmitting.

Figure 32 shows a part of the main screen that includes tick boxes for Auto Sequence, Call 1st and Hold Tx Frequency. Auto Sequence allows WSJT-X

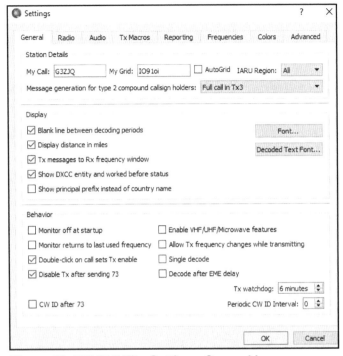

Figure 33: WSJT-X File-Settings-General box.

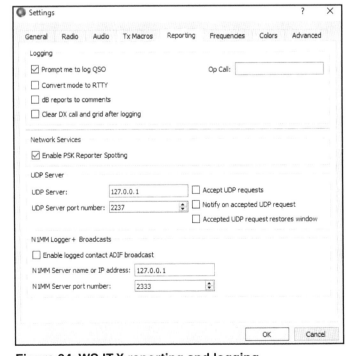

Figure 34: WSJT-X reporting and logging.

to transmit a series of standard messages that complete a QSO by exchanging callsigns, locators, signal strengths and finally a 73 message. Call 1st instructs WSJT to respond to the first decoded reply to your CQ call. Hold Tx frequency is self-explanatory and stops your transmit frequency from shifting as you respond to various stations.

You can alter the behaviour of the WSJT in the File-Settings-General box as shown in **Figure 33**. It is convenient to let WSJT set the Tx enable as well as populate the standard messages with the callsign of the station you are trying to contact when you double click on the stations message in the Band activity window. This makes it easy to respond to a calling station. Similarly, it is also convenient to let the software disable the Tx after the contact is completed.

Finally, it is useful to setup a prompt to yourself to log each contact and also to setup reporting across the internet assuming that you have internet access from the computer running the WSJT software. **Figure 34.** shows how you can do this by enabling PSK Reporter Spotting in the File-Settings-Reporting tab.

By now you should be eager to use the software and make a call but first 'listen' to the band and make sure your intended

3: Operating digital modes

transmit frequency is still unoccupied. If it is, click on the button box next to the CQ message Tx 6 and enable the Tx. You should see your transceiver go to transmit in the next available time slot and the waterfall will freeze. The CQ message is sent again after listening for a response and this will repeat until you either get a response, shown in red highlight in the Rx window, or you disable Tx enable. When you get a response, you just double click on it and the software will add the required information to the standard messages and begin the auto sequence. This whole process is shown in **Figure 35**.

Answering a CQ call (**Figure 36**) is just as simple. Check that your transmit frequency is still clear and just double click on the CQ message. The software will generate the standard messages, begin transmitting on the next time slot and complete the QSO automatically!

Having enjoyed several contacts or at least having sent CQ messages, you can monitor your performance using PSK Reporter found at https://pskreporter.info/pskmap.html. A typical map is shown in **Figure 37**.

Each callout in Figure 37 is a monitoring station. If you setup PSK Reporter Spotting previously, your own station will appear as one of these callouts. The lines connect your station to all those who have received your signals, so you can see at a glance how well your station is performing. If you rest your cursor on top of one of these monitors, then you will get some further information about the monitoring station and if this monitor has received your signal

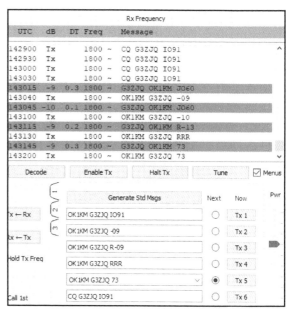

Figure 35. Originating a CQ call.

Figure 36. Answering a CQ call.

Figure 37: PSK Reporter map.

31

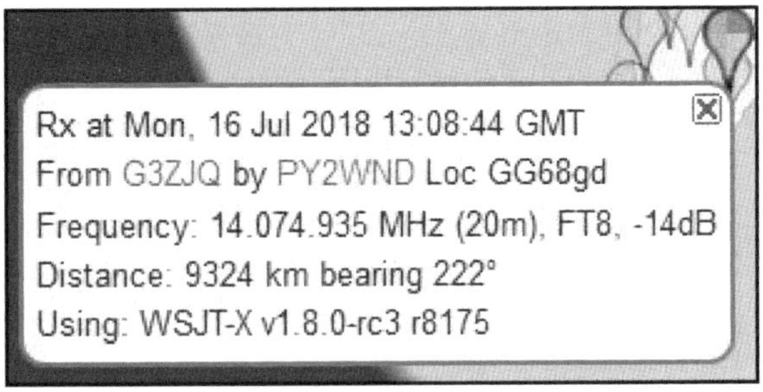

this will include a signal report. So far we have discussed operating FT8 mode in a normal amateur QSO context: a one to one communication between amateurs. However, one of the big advantages of digital modes – and FT8 in particular – is their weak signal capability and this makes them well adapted to DX operation. Even stations with relatively low power and modest antennas can make significant DX contacts using FT8. The two areas this has potential impact on are DX expeditions and contests. However, FT8 was until recently not well suited to the requirements of these activities. You may have noticed that a FT8 QSO takes several exchanges each one of which will take at least 15 seconds. This limits the QSO rate and makes operating pile-ups difficult to manage. The fixed message format of FT8 makes contest operating impossible when further information in addition to the callsign, report and locator is required. Version 2.0 of WSJT-X addresses both of these needs. **Figure 38** shows such a report from PY2WND for my 8 watt FT8 transmission using a half-size G5RV at 5 metres above ground level. FT8 is an excellent weak signal mode!

```
Rx at Mon, 16 Jul 2018 13:08:44 GMT
From G3ZJQ by PY2WND Loc GG68gd
Frequency: 14.074.935 MHz (20m), FT8, -14dB
Distance: 9324 km bearing 222°
Using: WSJT-X v1.8.0-rc3 r8175
```

Figure 38: PSK Reporter signal report.

3: Operating digital modes

FT8 DXpedition operating mode: Fox and Hounds

A new FT8 operating mode, first introduced in WSJT-X 1.9, is aimed at DXpeditions and it is called Fox and Hounds mode. The aim is to allow the DXpedition, the Fox, to complete many more QSOs per hour with the Hounds than the normal FT8 operating mode. This is achieved by allocating the first 1000Hz of the FT8 waterfall exclusively to the remote DXpedition or Fox. The region between 1000Hz and 4000Hz is where the calling stations, the Hounds, call the Fox. By separating the two it allows the Fox to respond to several calling stations during each 15sec time slot and hence improve the QSO rate. The operating sequence is partly automated and follows the following stages.

1. The Fox calls CQ within the 300-900Hz region
2. Several Hounds respond in the next time slot using frequencies above 1000Hz
3. The Fox responds with signal reports to each station again within the 300-900Hz region
4. Each Hound completes the exchange by giving the Fox his report on the same frequency as he was called upon by the Fox. This frequency shift is controlled by the software.
5. The Fox completes each QSO with a RR73 message

To operate in this new mode as a Hound you will need to set up CAT control of your transceiver, split frequency operation (both these setting can be found in File-Settings-Radio), check *Monitor returns to last used frequency* (File-Settings-General) and select Hound in the Special Activities sub-section of File-Settings-Advanced: see **Figure 39**. Since this process is quite different to the normal QSO you should not attempt to contact the Fox using the standard FT8 mode.

It is worth noting that the default Hound mode does not decode signals above 1000Hz but they are still vis-

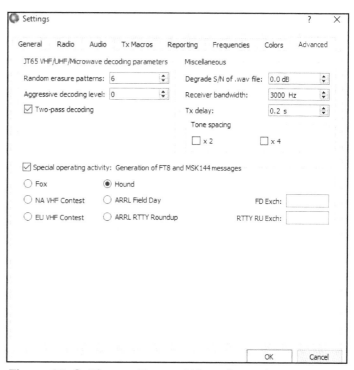

Figure 39: Setting up Fox and Hounds mode.

33

```
    Fox                              Hounds
1.  CQ KH1/KH7Z
2.                                   KH7Z K1ABC FN42, KH7Z W9XYZ EN37, ...
3.  K1ABC KH7Z -13
4.                                   KH7Z K1ABC R-11
5.  K1ABC RR73; W9XYZ <KH1/KH7Z> -17
6.                                   KH7Z W9XYZ R-16
7.  W9XYZ RR73; G4AAA <KH1/KH7Z> -09
8.  ...
```

Figure 40: Fox and Hounds QSO exchange.
(http://physics.princeton.edu/pulsar/k1jt/FT8_DXpedition_Mode.pdf)

ible on the waterfall so you can find a clear frequency. If you want to monitor the pile-up you can decode all received signals by checking the *Rx All Freq* box on the main window. Some options used in the standard mode will disappear from the main window when in Hound mode and a bright red box indicating Hound mode will appear. Good operating practice is not to call the Fox if:

1. You cannot hear him;
2. The Fox is call using a directed CQ and you are not within that continent.

To call the Fox you double click his decoded message. A typical exchange is shown in **Figure 40**.

The Fox, KH1/KH7Z, calls CQ and several stations respond. The Fox replies to K1ABC with his signal report. Next K1ABC sends a signal report to the Fox. Fox then confirms the QSO with KIABC and sends a signal report to W9XYZ in the same time slot. The ability of the Fox to send multiple messages within the same time slot increases the QSO rate over the standard mode. There are no RRR or 73 messages from the Hound and this also helps to keep the QSO rate high.

Operating as the Fox is slightly more complex since there are more things to manage. WSJT-X has several features in Fox mode that helps you manage the pile-up.

1. The left hand pane on the main window becomes a Stations calling Fox window which can be sorted by distance, signal strength, time (age) etc.
2. There is an additional log window.
3. The number of simultaneous transmissions can be increased allowing Fox to respond to several stations at the same time.

More detail about DXpedition mode can be found at:
https://physics.princeton.edu/pulsar/k1jt/FT8_DXpedition_Mode.pdf

3: Operating digital modes

FT8 Contest Modes

The new FT8 protocol allows more information to be passed in each exchange and this has made FT8 more contest friendly.

- NA VHF Contest mode allows for /R callsigns
- EU VHF Contest mode allows for /P callsigns
- ARRL Field Day mode allows for the exchange of Field Day operating classes required by the contest rules. This information is entered into the FD Each: box found on the File-Settings-Advanced page
- ARRL RTTY roundup mode is similar to the Field Day mode and exchanges the required signal report (RST) and State.

These modes can be setup on the Special Activities section of the File-Settings-Advanced page in WSJT-X in a similar manner to the Fox and Hounds mode discussed above.

Cooperating programs

The WSJT-X application is very powerful but there are some operating areas that it does not cover so there are other programs that can be used to fill in the gaps. One of the most popular is JTAlert. It provides several operating aids, including audio and visual alerts that track wanted DXCC countries, prefixes and grid squares, as well as automatic logging to several online logging programs. This program by VK3AMA is a free download available from http://hamapps.com/

It is only available for the Windows platform. The installation is easy using a standard setup file. You need to start WSJT-X first and then start JTAlert. Its main window is shown in **Figure 41**.

If you are receiving and decoding stations using WSJT-X, after each decode period, the stations heard will populate the spreadsheet like boxes in the main JTAlert window. Figure 41 shows six active stations. The alerts are colour coded and, in this case, the three stations in the darker boxes are calling CQ. The plain boxes contain stations that have no alert status and are just exchanging QSO information. By double-clicking on the box, you can initiate a reply to that station just the same as if you click the station within WSJT-X. There are many more alert types available as can be seen

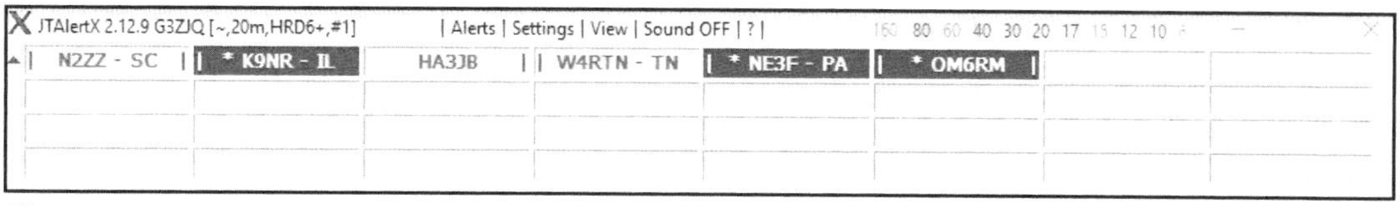

Figure 41: JTAlert Main window.

in **Figure 42**. Each alert is colour coded and these colours can be changed in the Alert types page as shown in **Figure 43**.

As you can see, JTAlert has many alerts and can quickly become confusing, so only turn on the alerts you find useful. The decode period for FT8 is only 15sec and, on a busy band, you can have twenty or more active stations, so these alerts can flash by very quickly. You can set an alert priority so that stations triggering several alerts will only show the colour of the highest priority alert. One alert that is very useful is the CQ alert, which has the default colour green but it is also indicated by an "*" character in front of the callsign (the hash character "#" indicates a directed CQ). If you use these alert indicators, rather than the colour alert, you can

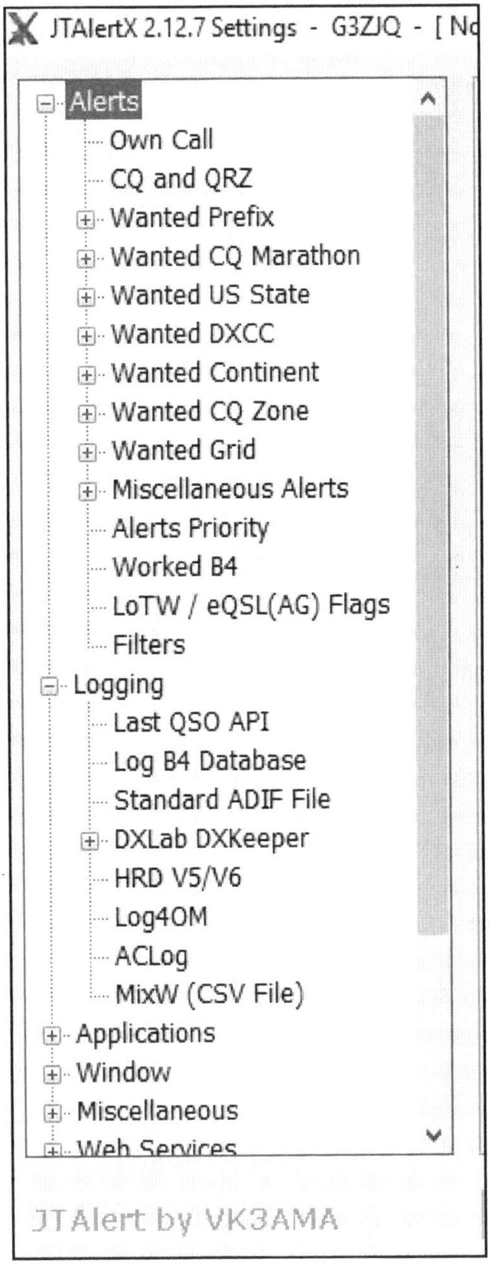

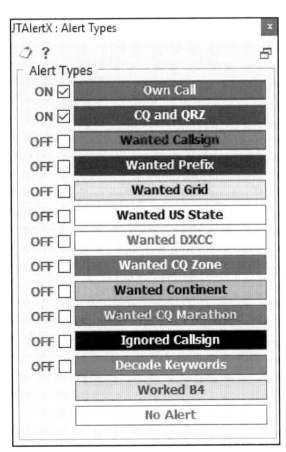

Figure 42: JTAlert settings.

Figure 43: JTAlert alert types.

3: Operating digital modes

combine the CQ alert with another alert, for example *Wanted DXCC*. It is then clear if the station calling CQ is on your wanted list: see LZ532PSO in **Figure 44** that shows a wanted DXCC calling CQ. Two other pieces of information are also shown by symbols rather than colours: membership to the Logbook of The World (LoTW) and membership of eQSL. LoTW membership is shown by a vertical stripe on the left of the box whilst eQSL is shown by a similar stripe on the right: see EW8W in Figure 44.

Figure 44: JTAlert boxes.

You can also setup audio alerts using .wav files that can be downloaded from the same site as the main program. If you wish to use audio alerts, make sure you set a different audio output device from the one used by WSJT-X, otherwise your alerts will be sent to your transceiver rather than your speakers! The audio output device is set in JTAlert via Settings-Sound Card.

JTAlert does more than just offer alerts, it also has a band monitor that indicates the number of active JT65, JT9 and FT8 stations on all of the bands between 160m and 6m: **Figure 45.** It can be made visible by selecting it in the View tab on the main window. This information is also found as a single row of coloured band reference numbers found at the top right of the main JTAlert window, see Figure 41. The colour code can be customised in Settings-Windows-Band Activity Reports.

Another useful feature of JTAlert is automated QSO logging. It is capable of working with several commonly used logging programs: DXKeeper, HRD, Log4OM and ACLog. It also allows automatic links to several online logbooks:

Band Activity

Unique Callsigns TX/RX per Band
Solar : SFI 71 : A 3 : K 1

	tx	rx	tx	rx	tx	rx		tx	rx	tx	rx	tx	rx
160m					76	26	17m					155	140
80m					220	104	15m					81	50
60m					69	48	12m						
40m	9	4	2	1	561	329	10m					6	6
30m					352	173	6m					35	11
20m	1	1			468	391	ALL	10	5	2	1	+1K	+1K
JT65		JT9			(Last Update : 23-Dec, 17:18 utc)								FT8

Figure 45: JTAlert Band monitor.

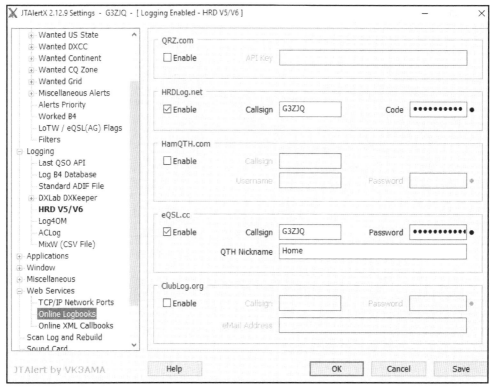

Figure 46: JTAlert logging options.

HRDLog.net and eQSL.cc. See **Figure 46**.

More details of these logging options can be found online using a simple search. More support for JTAlert can be found at
https://hamapps.groups.io/g/Support

I hope this section has demonstrated that you can set up a complete digital station around WSJT-X that is fully functional for DX chasing and Contests.

Using Fldigi in BPSK31 mode

Having completed the basic Fldigi configuration, you can explore the many different modes. Let's start with BPSK31. If you have followed the previous chapter on setting up Fldigi and you are tuned to an active PSK frequency the waterfall should look something like **Figure 47**.

Note that all of the signals that you can see will not necessarily be BPSK31 signals and you must get used to identifying the mode from either the sound or the shape of the signal shown in the waterfall: BPSK31 is shown in **Figure 48**.

It can take some experimentation to get the mode right, but you will quickly get to recognise the various modes. Fldigi calls the various decoders modems and provides you with the ability to alter the modem parameters but it is best

3: Operating digital modes

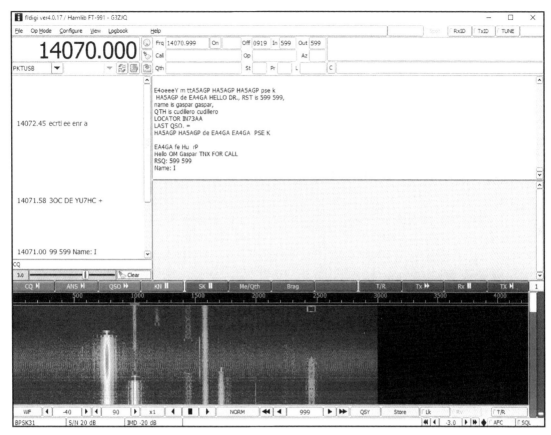

Figure 47: Fldigi in operation.

to leave these alone until you have gained more experience and knowledge.

The next stage is to adjust the Tx audio level to avoid overdriving your transceiver. This is done in a similar way to the setup for WSJT-X and, if you have set-up the Tx audio level for WSJT-X, it will be easier to set up Fldigi. Adjust the audio output level (speakers) on the sound card to midrange. In Windows 10 the level control is found in the same place as the microphone level adjustment explained earlier. Then adjust the mic or if using the rear input, the data input level on your transceiver so that your signal is clean. On some transceivers this can be done whilst watching the ALC level and adjusting the drive to the point just before the ALC comes into effect. However, a more accurate method is to get a friendly local BPSK31 operator to give you honest feedback on your signal when transmitting on a clear frequency away from the BPSK31 "watering holes". Another alternative is to monitor your own transmission using a Web SDR as you make the adjustments. Fldigi does offer an intermodulation

Figure 48: BPSK31 as it appears in the waterfall

39

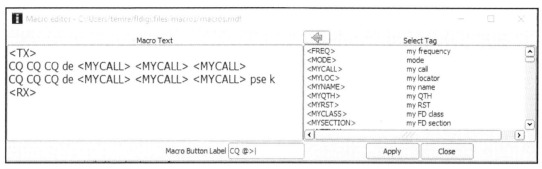

Figure 49: Macro editor.

distortion measurement of incoming signals as well as signal to noise ratio so you may be able to get an IMD report from another station.

If you type a message into the transmit window and toggle the Tx button, your transceiver into transmit and send the message on the frequency indicated by the red lines on the waterfall. It will not return to receive automatically so you must toggle back to receive using either T/R button, found on the command ribbon just above the waterfall, or by using Pause/Break key on your keyboard. On this command ribbon, you can find several macro keys. These will send defined text to the transmit window allowing you to send predetermined messages or parts of messages. For example, CQ is a standard message: CQ CQ CQ de G3ZJQ G3ZJQ G3ZJQ   CQ CQ CQ G3ZJQ G3ZJQ G3ZJQ pse k. If you right click the macro button it opens the macro editor with the macro loaded ready to be edited, **Figure 49.**

There are a number of set commands and parameters that you can use to create standard messages: <TX> acts like the PTT and turns on transmit and hence will send the following message without any further intervention, <MYCALL> is the callsign entered in the operator setup page, and finally <RX> will return to receive status. You can change the Macro Button Label to something you find more relevant, so you are not limited to the default ones. You can enter the information concerning the station you are contacting in the boxes at the top right, **Figure 50.**

If you enter the callsign and name of the person you are working, you can have these details automatically entered into any macro that uses the parameters <CALL> and <NAME>. You can also send the SN value in dB and the IMD automatically using <INFO1> and <INFO2> parameters in

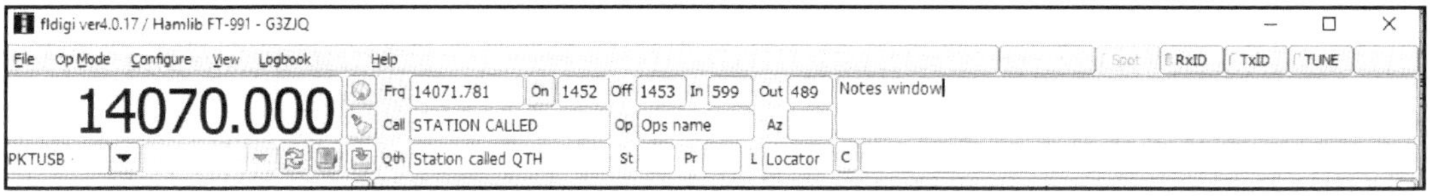

Figure 50: Entering the details of the station you are working.

3: Operating digital modes

your macro. The RST signal report is <RST> and needs to be entered in the Out box next to the notes window, see Figure 50. There are four sets of 12 macros to choose between. Finally, there are some keyboard commands worth noting: ESC will abort the current transmission and end it with an appropriate post amble, whereas hitting ESC three times will end the current transmission abruptly.

To select a frequency or respond to a CQ call you just click on the channel displayed in the Browser window and the red lines will jump to the appropriate frequency in the waterfall. The signal will be decoded and presented in both the Browser window and the Receive window. If you are not tuned to a valid BPSK31 signal, then you will see random letters appear on both of these windows.

Just as with WSJT-X, you can link Fldigi to PSK reporter by ticking the boxes under Configure->Miscellaneous->PSK Reporter.

Tips for using BPSK31

The following tips for using BPSK31 are from the ARRL website.
1. Use the centre of your audio passband whenever you can, since it offers the best Tx power and Rx capability.
2. Minimise the use of UPPER case letters. Lower case letters use less bits: see the later discussion of varicodes.
3. Use your transceiver's RF attenuator to avoid being desensitised by very strong signals
4. Keep your ALC reading as close to zero as possible
5. Remember, BPSK31 has an 80% duty cycle, so keep the power down to protect your equipment.

Summary

In this chapter we covered digital mode operating using FT8 and BPSK31 as example modes. Two software packages, WSJT-X and Fldigi were discussed in detail, including the important topic of setting-up the audio Tx drive levels to avoid generating harmonics.

4

Technical background to the digital modes

Introduction

The aim of this chapter is to cover the technical background to digital modes, to explain how they work and to give some reasons for their high performance in the weak signal regime of the DX QSO. We start with a primer on the relevant digital electronics, since this is barely covered in the average amateur enthusiasts training. This leads to an overview of a generic digital transmission system where the component parts are briefly explained. We then go into further detail of each one of these components. The mathematical background is kept to an absolute minimum and, where possible, diagrams are used in place of mathematical rigor. Examples from amateur digital modes are used where possible to link the theory to the practical amateur communications world.

Digital technology in the analogue world

Obviously, digital modes rely on digital techniques and that means we are in the realm of bits, bytes and codes. Linking these together is mathematics and software. It is well known that these are complex topics, and maybe beyond the average amateur to experiment with, but this is only partly true. They are complex but not beyond us amateurs. I will try and explain the magic behind these digital modes without the use of advanced mathematics.

The basis of digital systems is binary representation: everything is measured in bits; a bit is either '1' or '0' and groups of 8 bits are termed a byte. Long before digital systems entered everyday life, mathematics had studied the properties of bits and bytes in binary arithmetic. Of course, this is also the basis of computers and their software. Digital modes can be described as a fusion of communications and computers. Both of these areas also need to relate to us and we, of course, do not 'speak' binary so we need to link the digital domain to the real world of electronics, analogue voltages, and human language. This is achieved using analogue to digital converters and codes.

The English alphabet has been coded into binary many times using several codes, but the most significant one is the American Standard Code for Information Interchange or ASCII for short. This code gives each character a seven-bit code. For example, lower case 'a' is 1100001, whilst capital 'A' is 1000001 and the number '1' is 0110001. These binary numbers could be written as their decimal equivalents: 'a' = 97, 'A' = 65 and '1' is 49 but this is not normal practice. A more common technique is to use the hexadecimal numbering system. As the name suggests, this uses a base of sixteen, whereas our familiar decimal system uses a base of ten. For example, the ASCII letter 'a' is 97 when written in decimal, because it is comprised of nine 10s and seven 1s or units. In the hexadecimal numbering scheme, the same number would be written as 61. This is because it is made-up of six 16s and one 1 or unit. I'm sure you can see that there is potential for great confusion should decimal and hexadecimal numbers get intermixed. To avoid this, it is common practice to prefix a hexadecimal number when used in mixed numbering situations. The standard prefix is 0x so the letter 'a' in our example would be written as 0x61 in hexadecimal. One other point to note about hexadecimal is the characters we use. In decimal we use the digits 0 through to 9 but we need more to cope with hexadecimal. The solution is to extend the numbers with alphabetic characters. As a result, hexadecimal uses the following mix of numerical and alphabetic characters: 0, 1, 2, 3, 4, 5, 6, 7, 8, 9, a, b, c, d, e, f. Using this scheme, the decimal number 15 would be written as 0x0f in hexadecimal.

Whilst binary data is just a series of 0s and 1s, in computing, the data is represented by differing voltage levels. Many different voltage standards exist, for example, TTL logic levels are based on a 5v supply, whereas many of today's systems use much lower voltages. The lower voltage systems save power but are more susceptible to noise. In practical systems, it is not possible to have the voltage swing from exactly 0 to 5V to indicate logic 0 or 1, so a threshold is used. In a TTL system, a voltage higher than 2V is treated as a logic 1, whilst a signal lower than 0.8V is a logic 0. Similar thresholds are used for other logic systems to enable analogue voltage changes to be converted into a digital data stream.

There is one other dimension before we can form a data stream, and that is time. When do we make the threshold decisions? To do this, we introduce the concept of a digital clock. This is merely a square wave of constant period or frequency that determines the time at which we apply the thresholds to

4: Technical background to the digital modes

the voltage waveform. Now we can see how the input signal at the top in **Figure 51** is turned into the binary data stream below. This is termed a synchronous data stream because it uses a clock to synchronise the data.

In terms of serial data communications, we still have one more thing to achieve. If this data stream was a series of letters making up a message, we still need to know when each letter code starts and finishes. One way to do this is to make the start and finish of each letter code unique in some way. A simple way

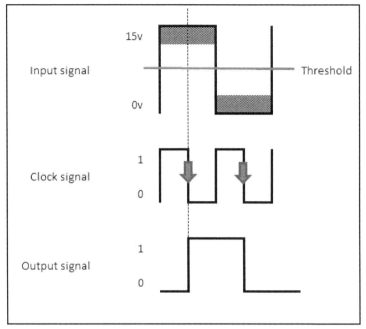

Figure 51: The clock signal controls the timing of a digital system.

to do this is to add a start and stop bit as shown in **Figure 52**. The start bit is a transition from high to low that tells the receiver to expect a data package, in this case a 4-bit sequence 1110 followed by a stop bit which is a transition from low to high. The key is that the timing between start and stop bits is known by the receiving system so it can synchronise the data by looking for this pattern.

As we will see later, adding more bits to the basic message is a common activity in serial data communications.

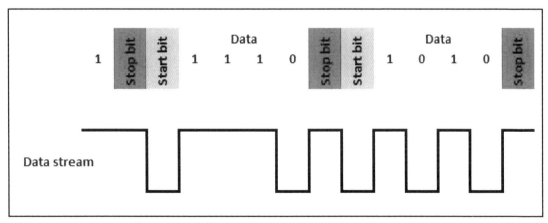

Figure 52: Start and Stop bits define the data.

45

How digital technology interfaces to the analogue world

The other aspect is linking the real world of analogue to the digital domain. An analogue to digital converter (ADC) and its inverse, a digital to analogue converter (DAC), are the items that do this task. First of all, let's consider a simple signal that should be familiar to all radio amateurs: the sine wave representing an AC voltage.

This could represent anything from AC mains through to an RF microwave signal. The key characteristics are amplitude and frequency (or period). That shown in **Figure 53** is a 100Hz 2 volt peak-to-peak waveform. To convert this into the digital domain, we measure the waveform at regular intervals of time – every T_s seconds. We term $1/T_s$ the sampling frequency or rate and each measurement is termed a sample. **Figure 54** shows an expanded version of the first part of Figure 53 with the sampling points marked. The table on the right-hand side of Figure 54 shows the samples taken every 500μs, which is a 2kHz sampling rate. The second column shows the voltage of the waveform at the time the sample was taken. The ADC converts this voltage to a binary number but, for convenience the third column is in decimal not binary.

In this example, the output values range between 512 and 1024, but the full range is 0 to 1024, with the values between 0 and 512 being used for the values -1 to 0 volts. The example in Figure 54, is a 10-bit ADC, which means it outputs binary values between 0 and 2^{10} (1024 in decimal notation). So, the sample at time 0.001 sec is 1100101101 (813 in decimal notation). These samples form a binary data stream that the ADC software will write directly into memory. Software that is used to directly control hardware, such

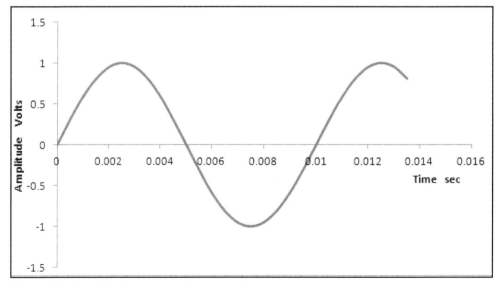

Figure 53. Amplitude versus time waveform.

4: Technical background to the digital modes

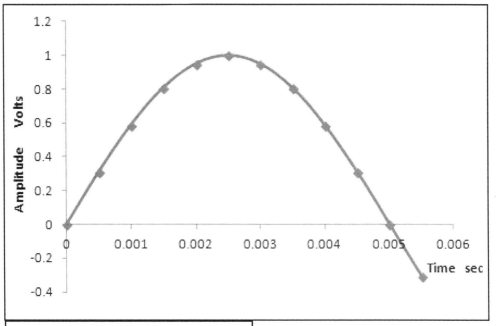

Figure 54: ADC converts a waveform to digits.

	Amplitude	
Time	Volts	ADC output
0	0.000	512
0.0005	0.309	670
0.001	0.588	813
0.0015	0.809	926
0.002	0.951	999
0.0025	1.000	1024
0.003	0.951	999
0.0035	0.809	926
0.004	0.588	813
0.0045	0.309	670
0.005	0.000	512
0.0055	-0.309	354

as the ADC, is known as firmware and usually resides in non-volatile memory as part of the ADC chip. It is becoming increasingly common to integrate much of the core functionality (ADC, controller and memory) into a single chip that can be used as a building block for more complex systems.

The typical modern PC sound card uses an audio codec that meets Intel's High Definition Audio standard. This includes a 16-bit ADC which can output 16-bit binary samples at a sampling rate of 48kHz. Software applications such as WSJT-X will store these values in memory ready for further processing.

The digital to analogue converter is the inverse of the ADC. It takes a binary value and produces the corresponding analogue voltage so that, given a stream of binary values, it can produce a waveform of voltage versus time. The frequency of this waveform depends upon the rate at which the samples are converted back to analogue voltages. This is often linked to an external digital clock that is the equivalent of the sampling rate of the ADC. By varying the clock frequency, you can alter the frequency of the output waveform. In Figure 54. the Time column in the table shows a steadily increasing time which in the case of the ADC was the sampling points. If, instead of sampling, we generate a clock with the same period and use this to clock a DAC whilst reading the binary values given in the 'ADC output' column, now held in memory say, into the DAC, we can generate the sine wave directly. It will appear at the DAC's output as a changing analogue voltage. This is the principle of direct frequency synthesis. In practice filters are required to clean up the output.

The PC sound card codec will also perform the DAC function at sample rates of up to 192kHz and sample depths of 8 to 32 bits. This is more than adequate for the generation of the tones used in HF digital modes.

Overview of a digital transmission system

A modern digital transmission system is shown in **Figure 55**. It shows the information flow through the individual stages that make up the system.

It begins at the top left with the user message that, in our terms, will consist of amateur callsigns, signal reports, location information, salutations and perhaps other station information or matters of a personal nature. However, there are two distinct categories of messages: structured messages that just contain the absolute minimum of information to establish a QSO and the more open-ended and freely structured message. The WSJT suite of digital modes caters for the former category of message, whilst the

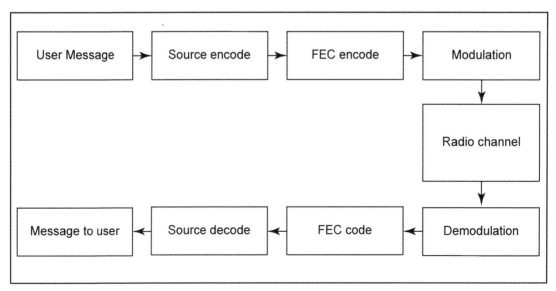

Figure 55: Information flow in a digital communication system.

4: Technical background to the digital modes

open-ended, sometimes called keyboard mode, messages are sent using RTTY and PSK31 modes.

Both message types benefit from some form of message compression to reduce the length of the message to be sent over the air and this is achieved using some form of encoding. WSJT modes use a very structured message that can compress the message into less than 75-bits. PSK31 uses a specific alphabet coding called varicode that also reduces the message length.

Having made the effort to reduce the number of information bits in the message, we now expand the message by introducing error control coding: forward error correction (FEC). This is a major topic in digital communications and information science and is responsible for the accurate transmission of digital data from storage devices, such as hard drives and DVDs, as well as satellites which provide us with HD TV. It is also vitally important in the communications over immense distances between the Earth and the many interplanetary probes. FEC is used in all of the WSJT modes as well as QPSK31 a variant of PSK31.

Digital communication uses a different modulation of the RF carrier from speech (analogue) signals. However, the two types of modulation are closely related and common digital modulation techniques, such as frequency shift keying (FSK), are easily achievable using amateur equipment.

The radio channel itself is, of course, the same for both analogue and digital transmission. Some of the annoying characteristics such as fading, noise, multipath distortion and Doppler shift become more important for the successful and accurate transmission of digital data. It is often these characteristics that determine the design of the other parts of the system: modulation technique and FEC for example.

Finally, at the receiving end of the communication system, we must demodulate and decode the signal to restore the original message. Sometimes this is a straightforward inverse of the encoding techniques, but often it involves different and quite complex methods. For example, in the case of FEC, the encoding and decoding often use different techniques.

Source encoding

The purpose of source encoding is to remove redundant information from the message before it is sent, thus making the process more efficient. If you have ever used file compression on your computer then you have used a form of source encoding. In amateur digital communications, two methods are common: one for structured messages like those in the WSJT modes and one for keyboard modes such as PSK31.

In the WSJT modes, the nature of the message to be sent is used to reduce the message size. Basic amateur QSOs follow a set pattern: exchange of callsigns, location, signal reports and some form of confirmation. Since all the world's callsigns confirm to a set pattern, and we already have a way of giving signal reports, this leads to a straightforward compression method. Callsigns consist on a one or two character prefix, which includes at least one letter, followed by a number and a suffix of one to three letters. Given the 26 letter alphabet and 10 numbers we have a total of 36 possible characters if we add the null character, then this makes 27 and 37 respectively. The number of all possible callsigns is then 37x36x10x27x27x27 which is over 262 million which is less than 2^{28}. Thus, we can code any callsign into 28bits and, in fact, more than 6 million of the possible callsigns are not needed and can be used for other messages such as CQ. The standard method of giving location is by using the Maidenhead grid locator, and there are 180x180 of these worldwide giving a total of 32,400, which is less than 2^{15} so we only need 15 bits to code the location. Thus, the standard QSO exchange of two callsigns, location or signal report or acknowledgement can be encoded into 28+28+15 = 71 bits. If we add three more bits to indicate alternative message structures then we have a total of 75 bits. This can be compared to the use of ASCII text that would use 100-136 bits, nearly twice as many bits. As we will see later, this would be expanded again when error control is added so any efficiency gains at this early stage affect the final message size dramatically.

FT8 employed the 75bit message format in all versions up to version 1.9 but in version 2.0 this changed to a 77bit message. This enabled some new features such as:
- Improved North American VHF contesting operation with full support to /R callsigns
- European VHF Contesting operation with six digit locator grids, QSO serial numbers and /P callsigns
- ARRL Field Day operation with standard Field Day exchanges
- Improved support for non-standard callsigns
- A new message format to allow the exchange of 71 bits of arbitrary information

Changing the message source encoding is not backwards compatible so users of the older versions must upgrade. Since this is free software it should not impose a burden on the user.

Such compression techniques cannot be directly applied to the keyboard modes such as PSK31, because, by their very nature, messages sent directly from a keyboard are less structured. However, we can make use of the frequency that each letter is used in a typical message to design the alphabetic code. Instead of using 7 bits for every character, as in the ASCII code, we could adjust the number of bits for each letter. In English, the most common letter is 'e' and the least common 'z'. So, if we use a '11' as the code for 'e' but '111010101' for 'z' we will be optimising the code for English. This is the basis of varicode alphabets.

4: Technical background to the digital modes

A recent development has created a hybrid keyboard mode, JS8call, which breaks free text messages down into blocks that can be sent using FT8 source encoding, but this is still in development.

Error detection

The human ear and brain are very good at deciphering information that is mixed with noise and even other sounds and conversations. It is this skill that CW operators put to great effect when they drag weak DX stations out from the chaos of an HF contest pile-up. A few mistakes can be tolerated and often common sense can fill in the missing information caused by erroneous reception. However, digital messages are far less forgiving of errors and techniques to improve the accurate communication of digital data now forms a major branch of communications theory. It is nice to know that radio amateurs are making full use of this theory and no doubt will continue to contribute to its development in the future.

When conversing with someone, errors or miscommunications are easily handled simply by asking the person to repeat the part we missed. However, life is more complex with digital messages because we first need to be able to detect that an error has occurred. One simple way is to use what's known as a parity check and there are two variations, odd or even parity. To use even parity checking, each character of the message being sent is examined to count the total number of logic 1s, and an extra bit is then added and set to either 1 or 0 to make an even number of logic 1s. At the receiving end, each character is examined again, and those with an even number of 1s are accepted, whilst the rest are assumed to be damaged and are rejected. Odd parity uses the same technique, except the extra bit is used to create an odd number of 1s. Parity checking is simple to implement with a low overhead but can easily be fooled if several bits have been corrupted. An extension of the parity error detection method is the cyclic redundancy check (CRC). This is slightly more complex but offers the ability to detect multiple errors at the expense of adding more bits to the original message. It is normally used on larger blocks of data than the parity technique. One of the key requirements of any error checking method is that it must be capable of being implemented using logic blocks that are commonly available in electronic circuits. The CRC calculation is based upon the exclusive OR function, which returns a 1 only if the inputs are different and is set out similar to long division. It is easier to understand if you follow the example in **Figure 56** (overleaf).

The message is padded out with zeros according to the length of the generator that in this case, is 10110. This generator is known to both the sender and the recipient of the message and is often described in the protocol, if this is a commercial system, for example CCIT X-25 uses 1000100000010001. The padded-out message is then divided by the generator using a particular form of division termed modulo-2 that is easily implemented using XOR logic

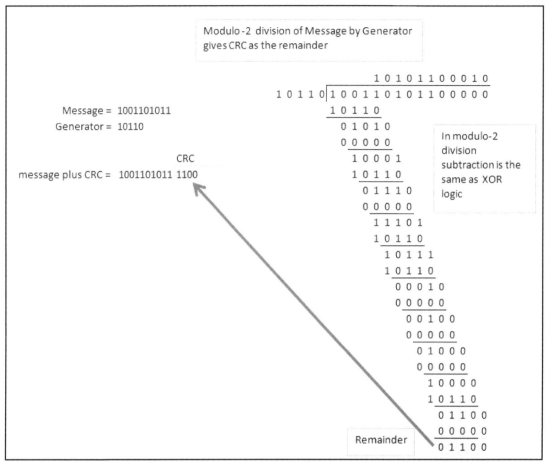

Figure 56: Example CRC calculation.

gates. The remainder from this process is appended to the message as the CRC. It is then easy for the received message to be checked – the receiver just carries out the same calculation on the message part of the received data and compares its calculated CRC to the received CRC. If they are the same, then the message is deemed error free. If not, then the errors have been detected but not identified.

Both parity and CRC are similar methods and they only allow you detect errors and hence confirm the message integrity. They are often combined with a method to request the sender to repeat the erroneous message in a hope that it will be received correctly the next time. These methods are termed automatic request (ARQ) methods. ARQ is used in amateur teletype over radio (AMTOR) mode.

4: Technical background to the digital modes

Forward Error Correction

Whilst it is useful to know if the message has errors it, would be better if we could also correct the errors in the received data. This is the aim of forward error correction (FEC). This has been the focus of much research as digital communications have become ever more important. FEC is used in controlling the errors in the transmission of data from DVDs, hard drives, tapes as well as satellite TV and DAB radio. It is a large and, at times, complex topic that is based upon relatively obscure mathematics; however, it is possible to grasp the basics without recourse to too much maths.

The basic FEC concept is counter-intuitive. To reduce the likelihood of errors in a message, we increase the message length, turning say a 12-bit message into 120-bits! How can this possibly work? Surely, the likelihood of an error increases with the length of the message? This becomes even more surprising if we keep the transmit time of the message the same as for the original message. With the transmitted energy the same for both, the critical measure of energy per bit is reduced, thus making the signal to noise ratio worse. However, this technique does really work, and it is the basis of all FEC.

Consider a simple example, we wish to send a simple message consisting of ASCII characters and we know from our previous discussion that these characters are represented as 7 bit binary numbers. The first four letters are shown in **Figure 57** where we can see that the letter 'b' is represented by 1100010_b in binary whilst the latter 'a' is 1100001_b. We can see that the difference between these two representations is the last two digits: 'a' ends in 01_b whilst 'b' ends in 10_b. This difference can be measured not only as an increase of 1, $01_b=1$ whilst $10_b=2$, but also as the number of changes that are required to change 01_b to 10_b which is two changes. This number of changes to convert one binary number to another is called the Hamming distance and it is a useful measure when we are considering errors because, errors are just that, changes from the original code. Figure 57 also shows the Hamming distance between the first four letters.

			Hamming distance			
	ASCI code		1100001	1100010	1100011	1100100
Letter	Decimal	binary				
a	97	1100001	0	2	1	1
b	98	1100010	2	0	1	2
c	99	1100011	1	1	0	3
d	100	1100100	2	1	3	0

Figure 57: Hamming distance between ASCI characters.

As you can see, the Hamming distance is small. In fact, just changing a single bit to the opposite will change the letter code to another character. This is made clear in **Figure 58** that shows the effect of a single error on each of the bits in the character 'b'.

This single error can change the received character from 'b' to 'r', 'j', 'f' etc. If we now change the representation of the characters from a 7 bit representation to a larger one, say 20 bits we can do the same analysis again.

The 20-bit codewords that represent the ASCII characters are, for this example, chosen at random. **Figure 59** shows the Hamming distance between these random codewords and, as you can see, it is much larger than the original 7-bit codes. So how does this help? Well, let us imagine that we receive a 20 bit codeword that has several errors. We were hoping to receive the 20 bit codeword for 'b', which is

Actual value	
1100010	b
Single bit error	
100010	"
1000010	B
1110010	r
1101010	j
1100110	f
1100000	'
1100011	c

Figure 58: Single bit errors.

10011000010000101011

but we actually received:

100**0**100**1**010**1**00101011, where the error bits are highlighted.

Figure 60 shows the Hamming distance calculation for this erroneous codeword measured against the four 20-bit codewords for the letters 'a' through to 'd'. As you can see, the minimum Hamming distance between the received codeword and the true letter codewords occurs with the letter 'b'. If our decoder used the minimum Hamming distance to select the correct letter, it would se-

Hamming distance		
20 bit codeword	Letter	100**0**100**1**010**1**00101011
11011001001001100101	a	9
10011000010000101011	b	3
11110100011000011101	c	12
10110010110010111100	d	12

Figure 59: Hamming distance for 20-bit codewords.

Hamming distance					
20 bit codeword	Letter	a	b	c	d
11011001001001100101	a	0	8	9	13
10011000010000101011	b	8	0	9	9
11110100011000011101	c	9	9	0	8
10110010110010111100	d	13	9	8	0

Figure 60: Codeword with 3 errors Hamming distance.

4: Technical background to the digital modes

lect 'b' since it is the nearest, with a Hamming distance of 3. Even though there were three errors in this received codeword, the correct letter was chosen. This is a significant improvement over Figure 58 where a single bit error in the message results in the incorrect message being received.

You might argue that, increasing the message length, would also increase the likelihood of more errors and that this technique would therefore not work. In fact, this was the view of the majority until 1948 when Claude Shannon completed his theoretical work to prove the benefits of an increased message length. There are some limitations; a key one being that this technique only works if the signal to noise level is above a certain limit: the Shannon limit. Above this limit, we can apply this technique of increasing the message length to reduce the received message error rate to an arbitrarily low level and, in theory, reduce the error rate to zero. Of course, in practice, this represents a target rather than a reality.

Although this seems counter intuitive, a simple explanation of how it works is possible. The key is the distance between the larger codewords, measured in our example by the Hamming distance. The larger this distance, the more unique each codeword is and the more it has to be distorted by errors to be confused with another codeword. In essence, we are averaging the errors over the longer codeword, rather than the shorter message and making use of the fact that the longer codewords are more distinct from each other. Whilst Shannon's theory tells us that this will reduce the message error rate, it does not tell us how to select the larger codewords. In this example we used random 20-bit codewords and these were not optimised at all, but we could still see an advantage. The practical difficulty is that randomly selected larger codewords are only possible in the simplest of cases. A 7-bit message has only 128 possible values but if the message was 75-bits long, then the number of possible messages would be $2^{75} > 10^{22,}$ which is far too many to be searched for, even using modern computers. We need a method to code and decode these longer messages in real time to take advantage of this error correction technique.

Shannon's work in 1948 did not provide a method to determine how you could generate codewords from messages or vice-versa, but others soon tackled the problem and today we have many methods available. These methods are based on a branch of mathematics called abstract algebra and are often named after mathematical concepts or the original inventors. Thus, we have convolution methods, Reed-Soloman codes (RS) and Low Density Parity Codes (LDPC), to name a few.

Convolution encoding

One of the earliest, and hence well studied, error coding techniques is convolution encoding. The term convolution refers to a mathematical technique, but

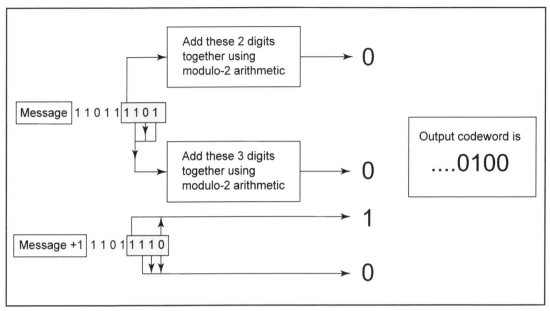

Figure 61: Convolution coding.

we don't need the maths to have a basic understanding of the coding process. Convolution coding can be applied to blocks of data, but it also works well with a data stream. The basic principle of convolution coding is similar to the CRC technique discussed, as it uses modulo-2 arithmetic but, unlike the CRC, it is best explained using a data stream rather than a block of data. It is illustrated in **Figure 61**.

The message is shown on the left-hand side of the figure and consists of the data stream 110111101 which is passed through a 4-bit wide window. The values within this window are combined in specific ways, determined by what is called a polynomial and uses modulo-2 arithmetic similar to CRC, to produce two output values per bit in the message. In the figure, the message is advanced by only 2 bits but the output shown on the right consists of 4 values, thus we are doubling the number of bits in the coded message. The width of the window, termed the code constraint length, is part of the convolution code's design as is the exact algorithms or polynomials to derive the two (or more) output parity bits that form the codeword.

Decoding the codewords is more complex than just applying the same algorithm at the receiver, since the actual message is not sent, just the output parity bits. The decoding is achieved using an algorithm first proposed by Andrew Viterbi in 1967 and hence the decoder is called a Viterbi decoder, and can be implemented in hardware or software.

Several of the WSJT modes make use of convolution coding and soft-decision decoding, including JT4, JT9 and WSPR. They use a constraint length of 32 bits and produce 2 parity bits per message bit. The polynomials

4: Technical background to the digital modes

used to produce the output are given in the WSJT User Guide but, of course, to use the WSJT modes you do not need to consider these at all. Just use the software provided!

Reed Solomon encoding

As you might expect, convolution coding performance is not the most effective error coding technique although it is very widely used in current commercial digital communications. It also forms the basis of other, higher performance, encoding techniques called Turbo encoding. Another very commonly used error coding technique is Reed-Solomon (RS) encoding. It is widely used in data storage devices – hard drives, DVDs etc. It is also used in digital satellite TV – DVB. In amateur circles, it is used in the WSJT mode JT65. RS encoding is a block encoding method which means it takes a block of data and encodes it by forming another longer block. RS codes are characterised by two numbers, n and k, k is the number of information bits to be encoded and n is the resulting number of bits in the encoded block. Thus in WSJT JT65, which uses RS(63,12), 12 information bits are encoded to become 63 bits. The JT65 message comprises 72 information bits and hence RS(63,12) encoding will produce a 6x63 = 378-bit coded message or codeword. Since the mathematics of RS encoding ensure that each codeword is separated from the next valid codeword by the maximum Hamming distance, good error correction is achieved. An illustration of this difference is shown in **Figure 62**.

```
Message #1:   G3LTF DL9KR JO40
Packed message, 6-bit symbols:   61 37 30 28   9 27 61 58 26   3 49 16
Channel symbols, including FEC:
    14 16   9 18   4 60 41 18 22 63 43   5 30 13 15   9 25 35 50 21   0
    36 17 42 33 35 39 22 25 39 46   3 47 39 55 23 61 25 58 47 16 38
    39 17   2 36   4 56   5 16 15 55 18 41   7 26 51 17 18 49 10 13 24

Message #2:   G3LTE DL9KR JO40
Packed message, 6-bit symbols:   61 37 30 28   5 27 61 58 26   3 49 16
Channel symbols, including FEC:
    20 34 19   5 36   6 30 15 22 20   3 62 57 59 19 56 17 35   2   9 41
    10 23 24 41 35 39 60 48 33 34 49 54 53 55 23 24 59   7   9 39 51
    23 17   2 12 49   6 46   7 61 49 18 41 50 16 40   8 45 55 45   7 24

Message #3:   G3LTF DL9KR JO41
Packed message, 6-bit symbols:   61 37 30 28   9 27 61 58 26   3 49 17
Channel symbols, including FEC:
    47 27 46 50 58 26 38 24 22   3 14 54 10 58 36 23 63 35 41 56 53
    62 11 49 14 35 39 60 40 44 15 45   7 44 55 23 12 49 39 11 18 36
    26 17   2   8 60 44 37   5 48 44 18 41 32 63   4 49 55 57 37 13 25
```

Figure 62: JT65 message encoded using RS(63,12) (Taylor, 2005).

Three messages are shown with a single character difference between them: G3LTF DL9KR JO40 and the two slightly different messages representing single differences or 'errors'

G3LT**E** DL9KR JO40 and G3LTF DL9KR JO4**1**.

Each message is shown in digital form by the row of numbers beginning with 61 37 .... It would have been better to show the true binary codes here but that would have made the figure very difficult to analyse so the binary numbers have been compressed into 6 bit values and written in decimal: the sequence 61 37 is really 111101 011111. These 6 bit values are termed symbols in the figure and we will meet this term later on when we discuss modulation methods. Notice that the three messages are very similar and in binary only differ by the fifth symbol or the last symbol corresponding to the position of the 'error' differences. Now look at the three encoded messages – they are completely different. This shows the advantage of using FEC encoding because these encoded messages are much easier to distinguish between even on a noisy radio channel.

Reed Solomon encoding occurs at the symbol level so the encoder will take the 12 symbol message – 61 37 30 28 9 27 61 58 26 3 49 16 and encode it to make the 63 symbols to be sent over the channel. RS(63,12) is capable of correcting up to 25 symbol errors in the 63 symbol codeword. Its performance correcting burst errors is thus excellent.

Low Density Parity Check encoding

In a search for encoding techniques that approach the Shannon limit, low density parity check (LDPC) encoding has been shown to be one of the best techniques and represents the current state of the art in FEC. Although they were discovered in the early 1960s their complexity and the lack of computing power at that time meant that they were ignored for more than 20 years. Although rediscovered in the 1980s they were not really developed any further until the 1990s. As computing power increased, this encoding technique has become more practical. Today LDPC encoding is seen as a potential FEC encoding technique in 5G mobile telephone networks as well as next generation Wi-Fi.

In amateur circles the popular WSJT mode FT8 uses LDPC FEC encoding combined with a CRC.

Synchronisation

So far we have produced our message, encoded it using an efficient source coding technique, added an error checking code, such as a CRC, and finally expanded the message length using an FEC technique. There is one more thing to do before we can modulate an RF carrier and transmit the message

4: Technical background to the digital modes

over a radio channel. We need to tell the receiver when the message starts! A common way of doing this is by using start and stop bits as was discussed in a previous section. This system has been in use by RTTY systems for over 50 years. To improve on this, PSK31 uses a varicode alphabet that has been designed to exclude the bit combination 00 – there are no characters in this varicode that contain the sequence 00. Synchronisation is then achieved by sending 00 between characters.

If the digital message is sent as a block message, then an alternative is to use a fixed synchronisation pattern that is added to the actual message before it is sent that will allow the receiver to synchronise with the transmitter. For example, JT65 interleaves a pseudo-random synchronisation pattern with the message. Synchronisation is so important that half of the final message is devoted to this sequence. FT8 uses three patterns, termed Costas arrays, one at the beginning, one in the middle and one at the end of each message block.

In highly structured messages, such as those produced by a WSJT mode, universal coordinated time UTC is used to allow the receiver to synchronise to the whole message sequence. It is for this reason that an accurate UTC clock is needed. However, this is not accurate enough to determine the exact message start, so some form of bit or symbol synchronisation method is still needed.

Modulation methods I - analogue

We are all familiar with the basic modulation types that have been in use by amateur stations for many years: amplitude modulation (AM), single sideband (SSB), frequency modulation (FM) and, of course, CW. Digital modes use these and even more exotic types, but the most important modulation type for HF digital modes is frequency shift keying (FSK) and phase shift keying (PSK). Before we discuss these in detail, let us back track to AM and SSB. In AM the radio carrier is amplitude modulated by the audio signal which is usually speech. **Figure 63** shows this can be

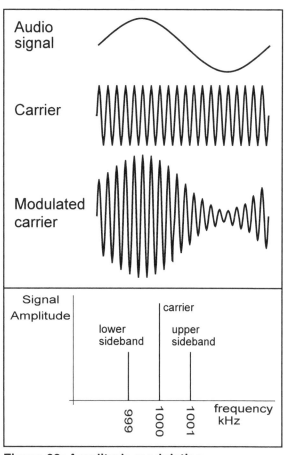

Figure 63: Amplitude modulation.

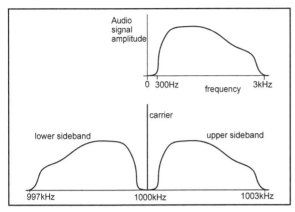

Figure 64: Single sideband modulation.

represented in two different ways: in terms of frequency or in terms of time.

Typically an AM signal occupies twice the audio bandwidth having upper (USB) and lower (LSB) sidebands centred on the radio carrier frequency. AM is quite easy to generate and can be received even on a crystal set. It was very common in the early days of radio, including amateur radio. Commercially, it is still used today on the long and medium wave bands and remains an option on most current HF amateur transceivers. For working DX using speech, AM has been replaced by SSB because this modulation technique is more efficient. This is because it puts all of the RF energy into the information carrying sideband and it has a smaller spectrum requirement.

There are several methods of creating SSB but the classical method is by filtering an AM signal in the IF stages to remove one sideband and the carrier. In a transceiver, this filtering is shared by both transmit and receive functions. Typically, this is a 2.7kHz bandwidth crystal filter that defines the IF bandwidth. On modern transceivers, this filtering is often achieved using DSP techniques, which also offer the advantage of a user definable variable bandwidth.

Finally, we have what is often described as the most efficient modulation type: CW. Here of course, we switch the RF carrier on and off using a Morse key to generate Morse code. It uses a very narrow spectrum, less than that used by SSB and is preferred by many as the ultimate DX mode. It could also be described as the first digital mode, since it is binary: on or off.

Even CW occupies a finite bandwidth due to the keying rate. Any abrupt transitions in the waveform due to keying will also extend the bandwidth and it is for this reason that some form of filtering is used to avoid these key clicks. In modern transceivers this is often adjustable using the menu system.

Since the early days of radio communications, it was noticed that the narrower the receiver bandwidth, the less noise was received, so this gave narrow modes an advantage when working DX. Hence, CW was deemed more noise efficient then SSB and SSB was better than AM. Another factor was the power efficiency of each mode. In AM the transmitted power is split into three, the carrier and two sidebands, but the information (that is the speech) is only in the sidebands and they are mirrors of each other. Thus SSB, which transmits only one sideband containing all the information, is much more power efficient than AM. CW also has this power efficiency advantage over AM.

4: Technical background to the digital modes

Despite these advantages, all three of these modulation methods suffer from the affects of noise. Any received noise is added to the signal and, if sufficiently large, will swamp the signal and make recovering the information impossible. For this reason, the signal to noise ratio is an important measure of the quality of the radio channel. At the threshold where the noise makes the recovery of the information impossible, it becomes a measure of the modulation efficiency. In amateur radio circles, the signal to noise ratio is usually defined in a 2.5kHz bandwidth as:

$$SNR = \frac{average\ signal\ power}{average\ noise\ power\ in\ 2.5kHz}$$

It is usually expressed in dB rather than as a fraction.

$$SNR_{db} = 10log_{10}(\text{SNR})$$

As an example, it has been suggested that to resolve SSB requires a SNR of +10dB whereas CW can be resolved down to a SNR -18dB (http://www.pa3fwm.nl/technotes/tn09b.html). These figures are of course approximate and should only be taken as indicative of the difference between CW and SSB.

Before moving on to the modulation methods used in digital transmissions, we should briefly discuss another common modulation method, frequency modulation (FM). AM, SSB and CW are all amplitude modulation methods but we can also modulate the RF carrier's frequency rather than its amplitude. This is shown in **Figure 65**.

The frequency spectrum of an FM carrier has more sidebands than an AM carrier and hence the bandwidth requirement is larger. It is for this reason that most FM channels are found in the VHF and above frequency bands. Commercial FM stations can occupy 75kHz but amateur usage it is often limited to 5kHz. The advantage of FM is its higher noise immunity due to being less susceptible to additive noise. This is because FM receivers only respond to frequency variations and not the amplitude variations associated with most radio noise. For this reason, a commercial FM station can have up to 20dB signal to noise advantage over an equivalent AM station

http://www.commsp.ee.ic.ac.uk/~kkleung/Communications2_2009/Lecture5.pdf

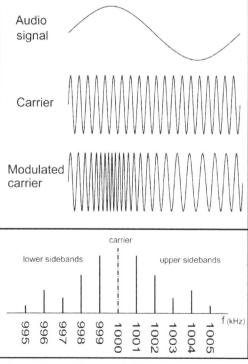

Figure 65: FM modulation.

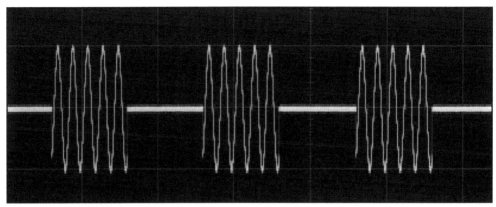

Figure 66: Amplitude Shift Keying.

Modulation methods II - digital

Digital modulation methods are an extension of the above modulation methods. As we discussed in the previous section, digital information is encoded into bits and bytes, so the modulation methods need to encode these bits and bytes onto an RF carrier in a similar way to the way speech modulation methods encode speech onto an RF carrier. The analogue modulation method that is similar to binary data is CW. **Figure 66** Amplitude Shift Keying, shows a carrier wave being switched on and off according to a binary signal.

Apart from the switching speed, each binary 1 represented by the carrier wave only contains a few cycles of the RF, this could be a Morse letter S albeit with horrendous keying harmonics due to the rapid switching. This form of modulation is called Amplitude Switched Keying (ASK). It suffers from the same susceptibility to noise as a Morse CW signal and, since digital transmission requires a low error rate, it is not used in practice except when combined with other modulation modes.

Following on from our discussion of the noise advantages of FM, we can consider the modulation mode shown in **Figure 67.**

Here the binary information is modulated onto a carrier by shifting the carrier frequency between two frequencies, one representing a binary 1 and the other a binary 0, hence the name binary FSK of BFSK. Figure 67 shows a simple 1010101 binary pattern modulation using two distinct frequencies. FSK, like FM, is less susceptible to noise than ASK. It is often the frequency shift that is specified rather than exact frequencies, for example, amateur Radioteletype (RTTY) is one of the oldest digital systems and it uses a frequency shift of 170Hz with the lower frequency representing a binary 1 and the upper frequency a binary 0. Other variants of RTTY exist that use alternative frequency shifts but the 170Hz shift is the most common.

It is not necessary to just use two frequencies for FSK. It is perfectly possible to use more frequencies. **Figure 68** shows a scheme using four frequencies.

4: Technical background to the digital modes

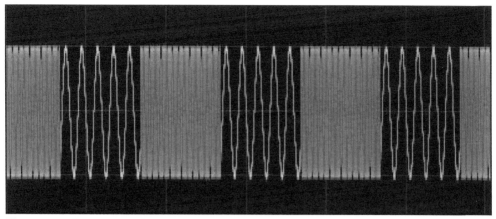

Figure 67: Frequency Shift Keying.

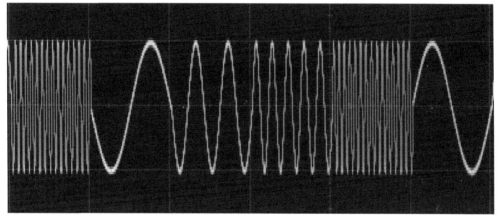

Figure 68: 4FSK.

However, the relationship to binary 1s and 0s then becomes slightly more complex now that we have four possible states. A simple encoding relates each frequency to not one binary digit (1 or 0) but to a sequence of two, for example, if the frequencies are F1, F2, F3 and F4 we could set F1=00, F2=01, F3=10 and F4=11. Now we have the added complexity of encoding a binary number, say 10011100, as a series of frequencies which in this case would be 10=F3, 01=F2, 11=F4 and 00=F1 so the sequence is F3 F2 F4 F1. To distinguish between binary (0 and 1) modulation and this, more complex form, we refer to the modulation elements as symbols, rather than binary digits. To make this notation consistent, we can define a modulation index that relates the size of binary number sent to the number of symbols in the modulation scheme.

$$Number\ of\ symbols = 2^m$$

Where m equals the number of information bits sent in each symbol. Thus in our example of 4 – FSK the number of symbols (frequencies) is 4 and

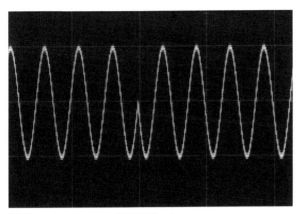

Figure 69: Phase Shift Keying.

hence m=2. Each symbol represents a two digit binary number. For example, FT8 uses 8 - FSK and hence m=3 for this mode: each symbol represents a 3 bit binary number. It is worth noting that, when each symbol represents more than one bit, there is a difference between the rate of transmission measured in bits/second and the rate of symbols transmitted. For example, if we are using FSK switching between two frequencies, as in RTTY, then each symbol represents just 1 bit and the bit rate is the same as the symbol rate but if we are using 8-FSK then each symbol represents 3 bits so the bit rate will be three times the symbol rate. The symbol rate is called the baud rate.

Whilst some form of FSK is used in the majority of the WSJT digital modes, there is another modulation method that is commonly used in amateur HF digital: Phase Shift Keying (PSK).

Figure 69 shows a binary PSK waveform (BPSK). Here the frequency of the carrier is not changed but the transition between a binary 1 and 0 is indicated by a 180 degree instantaneous phase change in the carrier. This is seen as the 'W' in the otherwise regular carrier sine wave. This can be generated in hardware by switching between two oscillators which are 180 degrees in phase apart or more often today by software using direct frequency synthesis within a PC sound card. In practice, such abrupt phase transitions will cause the signal bandwidth to be huge and some form of filtering is used in the same way as we remove key clicks in CW transmissions. In software this is done by smoothing the phase transition by imposing an envelope defined by a mathematical function: often a raised cosine curve. Demodulating BPSK requires an accurate measurement of the phase and which phase represents a binary 1 and a binary 0. This requires a very accurate receiver local oscillator and as such is not commonly used in amateur communications. One method to avoid this stringent requirement placed upon the receive local oscillator is to use differential BPSK (DBPSK). In this method the phase shift is with reference to the previous bit so that a series of 1s or 0s will not have any phase changes between each bit. There is only a phase change when there is a change from 0 to 1 or vice versa. DBPSK is used in the PSK31 HF digital mode developed by Peter Martinez in the 1990s. PSK31 is a high-performance mode that generally outperforms RTTY.

Just as with FSK, PSK is not limited to a single-phase transition of 180 degrees and multiple phase transitions are possible. Quadrature PSK (QPSK) uses four phases to encode the binary information and this is used in QPSK31

4: Technical background to the digital modes

a variant of PSK31. A similar modulation method called offset QPSK (OQPSK) is used in the WSJT mode MSK144 which is aimed at amateur meteor-scatter communications.

Just as we defined a SNR for analogue modulation, we can define a SNR applicable to digital modulation techniques. We replace the average signal power with the energy per bit E_b and the average noise in 2.5kHz by the noise power in 1Hz bandwidth N_0.

$$SNR = \frac{E_b}{N_0}$$

This is the preferred measurement of signal to noise ratio in the professional world and is used to measure the relative performance of digital systems. However, in the amateur world, we prefer the definition used previously for analogue signals, so the SNR provided by software packages such as WSJT-X use this earlier definition.

$$SNR_{db} = 10 \log_{10} \frac{average\ signal\ power}{average\ noise\ power\ in\ 2.5kHz}$$

Generating digital modes

Although these digital modulation methods seem complex they can be generated quite easily using an SSB transceiver and a PC sound card. The sound card can be programmed to produce any audio waveform using the direct synthesis method, where a series of numbers representing the amplitude of the waveform are sent to a DAC. FSK is achieved by creating a sequence of audio tones generated by the sound card and inputting these to an appropriate audio input on the transceiver. Sometimes this method of modulation is termed Audio Frequency Shift Keying (AFSK) but the result, when applied to an SSB transmitter, is identical to FSK.

If we input a pure audio tone, at say 1kHz, into a transceiver set to USB and tuned to 14.000MHz then the RF output will be on 14.001MHz. If we change the audio tone to 2kHz then the output changes to 14.002MHz. When sending an FT8 message we would be switching between eight tones whereas with JT65 it would be 65 different tones. It is this characteristic that gives each mode its recognisable warbling sound.

Generating, switching and streaming tones from the sound card is achieved in software using a specialist application programming interface (API) and although it requires quite intricate programming, it is no more difficult than any other programming activity. Of course, if you just want to use these modulation modes, then you do not need to program at all but just use the freely available software packages. However, if you want to delve deeper into the technology and understand how the software does what it does, then you will eventually meet APIs.

The radio channel

Radio channels are characterised by noise level, multi-path propagation, which causes fading and timing issues and Doppler shift due to movement somewhere in the path. These characteristics are generally random, although some can be expected to be present, their overall property is a lack of predictability.

Noise is present in all radio channels and it limits the ultimate communication performance of the channel. We usually distinguish between naturally occurring noise and manmade noise but the effects on performance are similar. What is more important are the properties of the noise. The basic noise floor is seen as a random fluctuation in received amplitude and has the characteristic of being relatively frequency independent. Hence it is termed broadband or white noise and represented by N_0, the noise power spectrum density, used in the calculation of the signal to noise ratio. In addition to this white noise, there are bursts of noise due to thunderstorms, lightning and, of course, manmade sources such as car ignition systems. The lower HF bands are very susceptible to lightning induced noise bursts, especially at night, whereas the higher HF and VHF bands are affected by ignition noise Power lines also produce a characteristic noise that adds to the overall noise present in the channel.

Multi-path reception occurs when there is more than one propagation path open in the channel. Various combinations of ground wave, single hop sky wave, line of sight, refraction and reflection can be present at the same time to make multiple paths. Different paths will have different path lengths and consequently, there will be timing and phase differences between the signals arriving at the reception point. This gives rise to interference between the signals causing fading and the different arrival times cause synchronisation problems for digital signals.

Doppler effects are due to relative motion between parts of the signal path. This could be due to movement of the source or the receiver relative to one another or reflections from a moving reflector, the moon, a meteorite or a satellite for example. It is not just the frequency that is shifted, the Doppler effect can also shift the phase of a signal. This can be very important for PSK modulated signals.

These characteristics of the radio channel form part of design parameters for the overall communication system. In analogue systems we rely on reducing the receiver bandwidth to reduce the noise and the human brain to decipher the signal as it fades in and out or is blocked by a noise burst. However, in digital systems, we must design the system to be as tolerant to the radio channel as possible. It is here that the coding techniques such as FEC and the modulation type and the rate of communication are all decided. For example, JT65 was originally designed for use in moonbounce communication, so it used a 46 second transmission period, a low communication rate of 2.7 baud, MFSK modulation that is tolerant to Doppler effects and a strong

4: Technical background to the digital modes

Reed Solomon FEC. Moon bounce is a slow communications channel. This can be compared to MSK144 designed for meteor scatter communications that uses a 72ms transmission period, a high communications rate of 2000 baud, OQPSK modulation and strong LDPC FEC. Clearly MSK144 is aimed at a fast-changing radio channel.

Demodulating the received signal

Tuning in to a CW or SSB transmission takes some care but with practice it becomes straightforward. Demodulating digital signals, on the other hand, used to require careful tuning to the exact frequency of the transmission before the decoder could operate. One of the many benefits of changing to computer-based demodulation of digital modes is the ability to demodulate and decode several digital signals at the same time. The waterfall display is now ubiquitous when using digital modes software, but how does it work? Understanding the display is easy, since it is just a visual representation what is being received with frequency across the bottom and time on the vertical axis. As you watch the display, it 'falls' down the screen recording the way in which the signals develop with time. It is a spectrum analyser that covers the 3kHz or so of the audio from the receiver. Until recently, computing this in real time was not possible, but with the increasing power of computers it has become available even on an average PC. In fact, many modern transceivers incorporate waterfall displays as part of their main frequency displays. The technology behind the waterfall is Fourier analysis.

Fourier analysis

Fourier analysis is generally accepted as an advanced mathematical technique that requires a strong mathematical background well beyond what the average radio amateur has achieved. However, the general concept is quite straightforward; it is a technique that calculates the frequency spectrum from an amplitude – time waveform. The frequency spectrum is a familiar concept to most radio amateurs since we often talk about filters, the bandwidth of signals and sidebands.

Figure 70 shows a typical frequency spectrum of

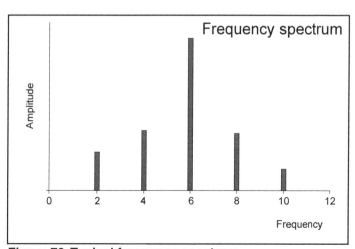

Figure 70. Typical frequency spectrum

67

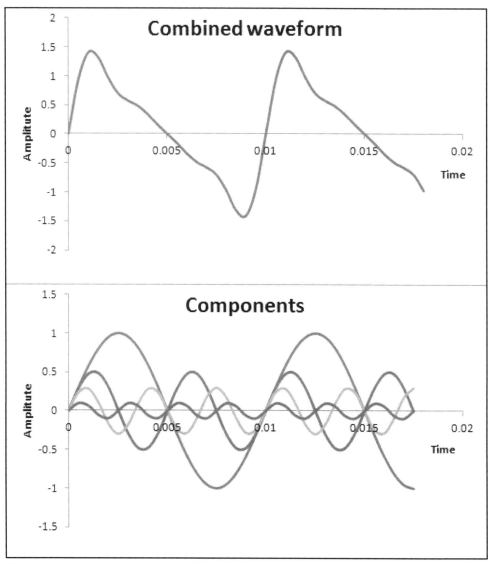

Figure 71: Combining frequencies.

amplitude versus frequency. It does not seem an easy task to convert the amplitude – time waveform to this spectrum; however, consider the opposite conversion. Can we convert a frequency spectrum to a waveform? Now all we need to do is to add up a series of time waveforms each corresponding to a specific frequency. **Figure 71** shows this process. Four waveforms of different frequencies are combined to make the top waveform. In this example, the frequencies are 100Hz, 200Hz, 300Hz and 400Hz and differing 'amounts' of each of these harmonics are added. This addition is point-wise, that is to say, each point on the four waveforms at a single time are added together

4: Technical background to the digital modes

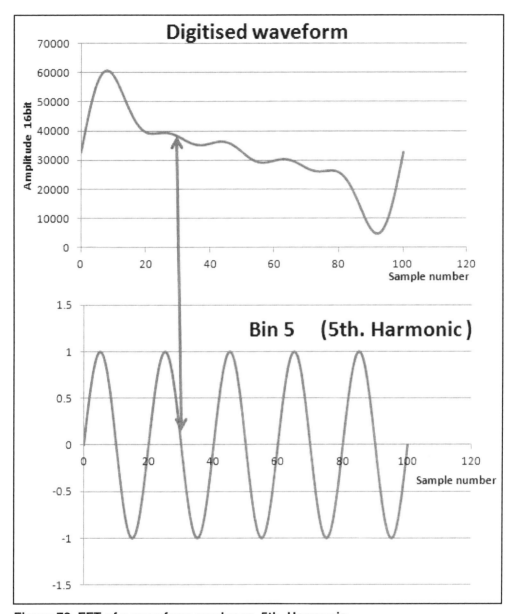

Figure 72. FFT of a waveform produces 5th. Harmonic.

to make the combined waveform. Notice that the combined waveform is not a pure sine wave but looks more like a saw tooth waveform.

The insight that Fourier had was that any amplitude – time waveform can be constructed in this manner from a series of pure sine waves whose frequencies are harmonically related. Of course, it often takes many, many pure sine waves to make any specific waveform. There is an assumption here that the combined waveform is repetitive – the shape of the waveform is repeated

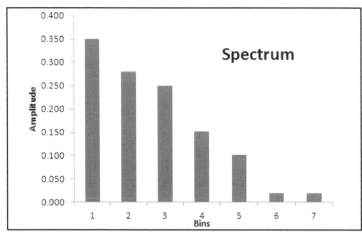

Figure 73: Amplitude of the first 7 FFT spectrum bins

over and over again. Figure 71: Combining frequencies. shows the repeating part of the waveform between 0 and 0.01 seconds.

When the waveforms are in digital / sampled form this combination process is straightforward. If each waveform is sampled in the same way you just need to add the binary values, something that is easy in software.

The reverse of this process begins by determining the periodicity of the input waveform. In practice this is usually achieved by mere selection: choosing to select 1024, 2048 or 4096 points from the waveform's digital stream and then repeating this process. There is a computational efficiency if you choose a power of 2 but this is beyond our present discussion. If we have a 48ksample/second data stream, then 4096 samples equates to approximately 85mS or 12Hz. Although choosing the number of points does determine the real temporal periodicity measured in seconds, time is usually measured in points rather than seconds.

Figure 72 (previous page) shows the digitised waveform to be analysed at the top. At the bottom is a pure sine wave that is a harmonic of the periodicity. Each harmonic is termed a bin and Figure 72 shows the calculation of the amplitude in bin 5. Each point in the digitised waveform is multiplied by the corresponding point in the lower pure sine wave and summed and averaged by dividing the sum by the number of points. In the case of Figure 72 this is a 100-point waveform, so we divide by 100. The amplitudes for the first 7 bins are plotted in **Figure 73**.

Bin **1** corresponds to a frequency of 100Hz, bin 2 to 200Hz and so forth.

If we have a 48ksample/second data stream, then 4096 samples equates to approximately 85ms or 12Hz so the FFT will provide a spectrum in multiples of this frequency: 12Hz, 24Hz, 36Hz and so forth.

In practice, the FFT calculation is repeated over and over to produce a moving spectrogram or waterfall display. This shows how the spectrum changes over time as illustrated in **Figure 74**.

The FFT is calculated four times and each bin has four levels corresponding to the value calculated for the bin at four consecutive times these are then transferred to a moving waterfall display where each value is assigned a colour. Each colour is assigned to a range of values. The waterfall display

4: Technical background to the digital modes

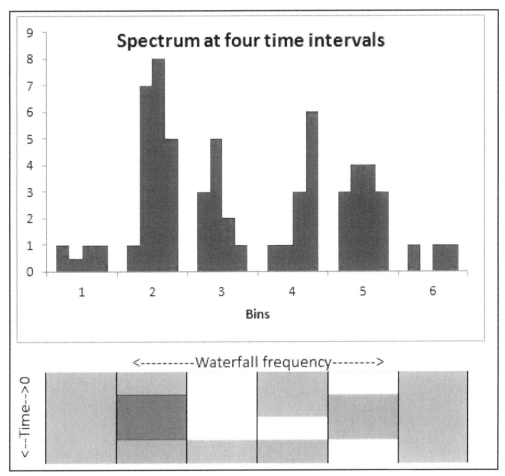

Figure 74: FFT and the waterfall.

is a key part of the tuning process in digital modes.

FSK is simple to demodulate in this way since the separate frequencies transmitted will fall in different frequency bins. It is then possible for the software to search the whole audio bandwidth and search out the FSK signals. Hence software such as WSJT-X can decode several digital QSOs at the same time and provide the operator with a list of stations that are operating.

PSK is slightly harder to demodulate and requires the receiver to synchronise with the carrier signal. However, the waterfall can be of use to select the signal from the ones that are available.

Source decoding

Source decoding poses few issues since the encoding process can simply be reversed to create the original message.

Summary

In this chapter we have covered the technology behind digital modes. Having first broken down a generic digital transmission system into its component parts, we then explored each part in depth. Particular attention was paid to the importance of error control and modulation method and how these combine in a well-designed system, to give the outstanding weak signal performance. Finally, we explained, using diagrams, Fourier analysis and the FFT, a mathematical technique that has revolutionised how we make use of digital modes.

5

The future of HF digital modes

In this book we have covered setting up a digital station, operating using HF digital modes and we have explored the underlying technologies that give these modes their high performance. We have used two popular software packages, WSJT-X and Fldigi, to illustrate the set-up procedures and two specific digital modes, FT8 and PSK31, to discuss operational techniques.

HF digital has become very popular in recent times. It offers better performance than SSB and may improve on CW. Despite this, general acceptance has not yet been achieved. There are some who feel that the automated text responses in modes such as FT8 are not in the spirit of amateur radio but change will always carry with it those who oppose it. Digital modes have a long history in amateur radio from the early days of RTTY to the present. Perhaps it is their progression into the main stream DXing and contesting community that has awakened the opposition by some.

Joe Taylor has recently updated FT8 in WSJT-X version 2.0 to include more features that are of interest to the contesting and DXing community. This major upgrade demonstrates that interest in weak signal modes is increasing and finding application outside of the technical domain and into main-stream amateur operating.

The keyboard modes, typically PSK31, have developed more slowly. There has been some development in the modulation techniques used moving from BPSK31 to other PSK modes that have more bits per symbol, such as QPSK. A recent development in keyboard-keyboard modes is JS8 which is an experimental mode based upon the FT8 protocol. A free application by KN4CRD that pioneers this mode is available online: search for JS8Call in groups.io. The software is available for Windows, Linux, MacOS and Raspberry Pi operating systems. The main window is shown in **Figure 75.**

If you have used either WSJT-X or Fldigi then this screen will look familiar. Since JS8Call is based upon WSJT-X, you will also find setting up this software, interface and your transceiver very easy. In fact, once installed, the software just needs your callsign and locator to be entered together with the audio devices you are using for digital modes. Your transceiver might not need any adjustment at all but it is worth checking the power levels, just in case!

As you would expect, operating using JS8Call is quite different to operating using FT8. To allow free format text, JS8 uses different source coding to FT8, so the two modes are not compatible with each other. The JS8 protocol breaks your text message down into 23 character blocks that it sends in back to back 15 second time slots using the FT8 error correction techniques, so it may take several minutes to send and receive even a short sentence. Nets and group communications are featured in JS8Call, including a novel feature that sends a message that all listening stations will automatically respond to giving you a signal report.

JS8Call is built upon the WSJT-X code base, which is only possible because WSJT-X is open-source software distributed under the GNU General Public License. The source code is freely available from SourceForge.net and the software tools needed to modify, compile and test the software are also freely available online. This makes it possible for anyone with the required skills to produce derivative software in a similar manner to constructing hardware to published designs, only cheaper! It is not necessary to do a complete project on your own, you can also contribute to the existing development teams.

Digital modes are not new to amateur radio, but they have been, until recently, a minority interest. What is certain is that, with modes such as FT8, they have entered the main stream interest and will continue to play a significant role in the future.

5: The future of HF digital modes

Figure 75: JS8Call main screen.

Another radiotoday guide

radiotoday guide to the Icom IC-7610

By Andrew Barron, ZL3DW

Within a few short months of the launch the Icom IC-7610 radio, it became a 'best seller' and one of the most popular radios on the market today. This SDR - based radio has simply become the measure by which other radios are being judged. However, getting the most from this radio, as it is for many new radios, is increasingly difficult. Andrew Barron, ZL3DW, an acknowledged SDR expert, sets out in this book to highlight the myriad of options available to the users of this fabulous radio.

On opening the box of the Icom IC-7610 not only will you be excited but perhaps you will surprised by how many settings and controls there are to learn. This should not deter you as this is where this book comes in. From the first steps with the panadapter display and its 'FIX' spectrum display mode where you can see the whole band or just a section of the band such as the CW segment, this book guides you. The biggest advantage that the IC-7610 offers over its rivals at a similar price level is the two completely independent receivers, so Andrew explains the changes this makes to the way you operate the radio. The touchscreen controls are explained so you get to know the radio through using it and through delving into every control and menu setting. There are tips on troubleshooting and guides to particular modes of operation and much more besides.

The IC-7610 is a truly exceptional radio and if you are interested in purchasing one or even already have one this guide provides invaluable reading. Many features are also applicable to other Icom SDR radios and the insight into the art of the possible with SDR radios is illuminating for all.

Size 174x240mm, 160 pages
ISBN 9781 9101 9366 2
Non Members' Price: £12.99
RSGB Members' Price: £11.04

Radio Society of Great Britain www.rsgbshop.org
3 Abbey Court, Priory Business Park, Bedford, MK44 3WH. Tel: 01234 832 700 Fax: 01234 831 496